Lecture Notes in Mathematics

1935

Editors:
J.-M. Morel, Cachan
F. Takens, Groningen
B. Teissier, Paris

Lecture Notes in Mathematics

André Unterberger

Alternative Pseudodifferential Analysis

With an Application to Modular Forms

 Springer

Author

André Unterberger

Mathématiques
Université de Reims
Moulin de la Housse, BP 1039
51687 Reims Cedex 2
France
andre.unterberger@univ-reims.fr

ISBN 978-3-540-77910-0   ISBN 978-3-540-77911-7 (eBook)
DOI 10.1007/978-3-540-77911-7

Lecture Notes in Mathematics ISSN print edition: 0075-8434
                              ISSN electronic edition: 1617-9692

Library of Congress Control Number: 2008921392

Mathematics Subject Classification (2000): 35S99, 22E70, 42A99, 11F11, 11F37, 81S30

Cover design: WMXDesign GmbH

Printed on acid-free paper

9 8 7 6 5 4 3 2 1

springer.com

To François Treves

Preface

The subject of the present work is pseudodifferential analysis: the motivations lie in harmonic analysis and modular form theory. So far as the last two domains are concerned, nothing more than some minimal familiarity is needed: some knowledge of the metaplectic representation, and of the definition of holomorphic and nonholomorphic modular forms, will help. Even though the symbolic calculus introduced here is entirely new, and does not depend on any technical result concerning pseudodifferential operators, it would not be honest to claim that no previous acquaintance with that field is necessary: the analysis developed here is strikingly different from the usual one, some knowledge of which – in particular, its representation-theoretic aspects – is needed for comparison.

Modular form theory is a very appealing subject: some time ago already, we tried to approach it from an angle which, to us, was much more familiar, that of pseudodifferential analysis. It is possible to realize nonholomorphic modular forms as distributions in the plane [35, Sect. 18], the main benefit being that they can then be considered as symbols for a calculus of the usual species, to wit the Weyl calculus. Yes, there are difficulties on the way toward developing the symbolic calculus of associated operators, since distributions on $\mathbb{R}^2$ which correspond to modular forms, though beautiful objects from the point of view of arithmetic, are extremely singular. Still, one can survive these difficulties, as shown in [36].

Only the nonholomorphic modular form theory could be reached in this way. Needless to say, we tried to incorporate holomorphic modular form theory as well: this cannot work to a full extent, and the best one can do in this direction will be summed up in Sect. 5.2 of the present work. Then, in an independent piece of work [38], partly motivated by Physics, we introduced the "new" anaplectic analysis – like many new things, it is only a coherent rearrangement of old ones – and it turned out, to our unanticipated satisfaction, that this solved our old problem.

Only one-dimensional anaplectic analysis will concern us here – the higher-dimensional case is considerably harder – and, of course, we are not assuming that the reader has read, or borrowed, our book on the subject. It is our opinion that the version presented here, in Sects. 2.2 and 4.1, in which no proofs are given, will make easy reading. Though our main current interest in anaplectic analysis lies with Physics, it is clear, to us, that the approach to holomorphic modular form theory it leads to deserves to be explored further.

Contents

1 Introduction 1

2 The Metaplectic and Anaplectic Representations 11
   2.1 The Metaplectic Representation 11
   2.2 Anaplectic Analysis 16

3 The One-Dimensional Alternative Pseudodifferential Analysis 27
   3.1 Ascending Pseudodifferential Analysis 28
   3.2 Classes of Operators 38
   3.3 The Resolvent of the Lowering Operator 57
   3.4 The Composition Formula 64

4 From Anaplectic Analysis to Usual Analysis 75
   4.1 The v-Anaplectic Representation 75
   4.2 Ascending Pseudodifferential Calculus in v-Anaplectic Analysis ... 83

5 Pseudodifferential Analysis and Modular Forms 93
   5.1 The Eisenstein, Theta, Poincaré, and Alternative Poincaré
       Distributions 93
   5.2 Moyal Brackets and Rankin–Cohen Brackets 106

References 115

Index 117

Subject Index 118

Chapter 1
Introduction

The object of the present work is to introduce a new "pseudodifferential" analysis on the real line. This alternative analysis is endowed with just as many symmetries as the usual one, mostly of a different nature. It is a necessary counterpart of the celebrated Weyl calculus, so far as certain matters pertaining to representation theory or modular form theory are concerned. Whether it may be of any use, in the foreseeable future, in partial differential equations, the most important domain of applications of pseudodifferential analysis, is much more questionable, though we shall try to give some minimal hints regarding related possibilities.

Though it is not our intention to discuss at length the well-understood harmonic–analytic aspects of the Weyl calculus, recalling some of its basic definitions and properties may be of help to some readers: also, it will facilitate the comprehension of the main features of alternative pseudodifferential analysis, which parallel the corresponding ones from the Weyl calculus but violently contrast with these most of the time. Finally, most of our readers are probably not familiar with the relation of the Weyl calculus to modular form theory, an important topic toward our present purpose.

One-dimensional pseudodifferential analysis starts with a way of representing linear operators acting on functions of one variable by functions of two variables. If $h \in S(\mathbb{R}^2)$, the operator $\mathrm{Op}(h)$, called the operator with Weyl symbol h [42], is the operator: $S'(\mathbb{R}) \to S(\mathbb{R})$ defined by the equation

$$(\mathrm{Op}(h)u)(x) = \int_{\mathbb{R}^2} h(\frac{x+y}{2}, \eta)\, u(y)\, e^{2i\pi(x-y)\eta} dy\, d\eta. \qquad (1.1.1)$$

It is easily seen (writing its integral kernel) that, under the sole assumption that $h \in L^2(\mathbb{R}^2)$, one can define the operator $\mathrm{Op}(h)$ as a Hilbert–Schmidt operator in $L^2(\mathbb{R})$. It is important, especially in the arithmetic (automorphic) case, to use the elementary fact that one can still define $\mathrm{Op}(\mathfrak{S})$ as a linear operator from $S(\mathbb{R})$ to $S'(\mathbb{R})$ if $\mathfrak{S}$ is an arbitrary distribution in $S'(\mathbb{R}^2)$.

As will be recalled in detail in Sect. 2.1, there is a unique (up to isomorphism) pair $(\widetilde{SL}(2,\mathbb{R}), \iota)$, where $\widetilde{SL}(2,\mathbb{R})$ is a connected Lie group and ι is a homomorphism from $\widetilde{SL}(2,\mathbb{R})$ onto $SL(2,\mathbb{R})$ with a kernel reducing to two elements. Moreover,

1

there is a certain unitary representation $\mathrm{Met}^{(1)}$, called the metaplectic representation, of $\widetilde{SL}(2,\mathbb{R})$ in the Hilbert space $L^2(\mathbb{R})$, one of the main features of which is that, for some well-chosen $\tilde{g} \in \widetilde{SL}(2,\mathbb{R})$ lying above the matrix $\begin{pmatrix} 0 & 1 \\ -1 & 0 \end{pmatrix}$, the operator $\mathrm{Met}^{(1)}(\tilde{g})$ is $e^{-\frac{i\pi}{4}}$ times the Fourier transformation. Up to the multiplication by ± 1, any operator $\mathrm{Met}^{(1)}(\tilde{g})$ with $\tilde{g} \in \widetilde{SL}(2,\mathbb{R})$ depends only on the image g of $\tilde{g}$ in $SL(2,\mathbb{R})$. All necessary details regarding this representation are to be found in Sect. 2.1. One then has the fundamental covariance formula

$$\mathrm{Met}^{(1)}(\tilde{g})\,\mathrm{Op}(\mathfrak{S})\,\mathrm{Met}^{(1)}(\tilde{g})^{-1} = \mathrm{Op}(\mathfrak{S} \circ g^{-1}): \qquad (1.1.2)$$

that the left-hand side makes sense depends on the fact that a transformation such as $\mathrm{Met}^{(1)}(\tilde{g})$ is also well defined as an automorphism of $\mathcal{S}(\mathbb{R})$ or as an automorphism of $\mathcal{S}'(\mathbb{R})$.

There is another covariance formula, actually involving the Heisenberg representation, which can be summed up in the following terms: if one sets $(\tau_{y,\eta}u)(x) = u(x-y)\,e^{2i\pi(x-\frac{y}{2})\eta}$, one has

$$\tau_{y,\eta}\,\mathrm{Op}(\mathfrak{S})\,\tau_{y,\eta}^{-1} = \mathrm{Op}((x,\xi) \mapsto \mathfrak{S}(x-y, \xi-\eta)) \qquad (1.1.3)$$

for every $(y,\eta) \in \mathbb{R}^2$.

These two covariance formulas lead to two quite different composition formulas. Given two symbols h_1, h_2 lying in $L^2(\mathbb{R}^2)$ (there is a considerable variety of other possibilities, which makes part of the tool kit of pseudodifferential analysis), the operator $\mathrm{Op}(h_1)\,\mathrm{Op}(h_2)$, as a Hilbert–Schmidt operator, can be written as $\mathrm{Op}(h_1 \# h_2)$ for a unique symbol $h_1 \# h_2 \in L^2(\mathbb{R}^2)$. The sharp composition of symbols can be analyzed in combination with the decomposition of functions in $\mathbb{R}^2$ under the representation involved on the right-hand side of (1.1.2) or (1.1.3). In the second case, we are dealing with the representation of the commutative group $\mathbb{R}^2$ acting by translations on $L^2(\mathbb{R}^2)$: the differential operators on $\mathbb{R}^2$ which commute with this action are the differential operators with constant coefficients, the joint (generalized) eigenfunctions of which are the functions $X = (x, \xi) \mapsto e^{2i\pi\langle A, X\rangle}$ with $A \in \mathbb{R}^2$. Then, the formula we are looking for is simply

$$e^{2i\pi\langle A, X\rangle}\,\#\,e^{2i\pi\langle B, X\rangle} = e^{i\pi\,[A,B]}\,e^{2i\pi\langle A+B, X\rangle}, \qquad (1.1.4)$$

where we have introduced the so-called symplectic form $[\,,\,]$ such that

$$[A, B] = -a\beta + b\alpha \qquad \text{if } A = (a, \alpha), B = (b, \beta). \qquad (1.1.5)$$

When making the operator $\mathrm{Op}(e^{2i\pi\langle A, X\rangle})$ explicit as $\tau_{a,\alpha}$, this formula is sometimes called the Weyl exponential version of Heisenberg's relation. More important to our purpose, when coupled with the decomposition of general symbols in $L^2(\mathbb{R}^2)$ into functions $X \mapsto e^{2i\pi\langle A, X\rangle}$ (this decomposition is nothing else than the standard Fourier transformation in $\mathbb{R}^2$), it leads to the quite well-known formula

$$(h_1 \# h_2)(X) = 4 \int_{\mathbb{R}^4} h_1(Y)\,h_2(Z)\,e^{-4i\pi\,[Y-X, Z-X]}\,dY\,dZ \qquad (1.1.6)$$

or, taking advantage of the unitary group generated (Stone's theorem) by the operator $\mathcal{L}$ on $L^2(\mathbb{R}^4)$ such that

$$i\pi\mathcal{L} = (4i\pi)^{-1} \left(-\frac{\partial^2}{\partial y \partial \zeta} + \frac{\partial^2}{\partial z \partial \eta} \right) \quad \text{if } (Y; Z) = ((y, \eta); (z, \zeta)), \quad (1.1.7)$$

to the fully equivalent formula

$$(h_1 \# h_2)(X) = \{e^{i\pi\mathcal{L}} (h_1(X+Y) h_2(X+Z))\} (Y = Z = 0). \quad (1.1.8)$$

Expanding the exponential into a series, one obtains the so-called Moyal formula

$$(h_1 \# h_2)(X)$$

$$\sim \sum_{n \geq 0} (4i\pi)^{-n} \sum_{j+k=n} \frac{(-1)^j}{j!k!} \left(\frac{\partial}{\partial x} \right)^j \left(\frac{\partial}{\partial \xi} \right)^k h_1(x, \xi) \left(\frac{\partial}{\partial x} \right)^k \left(\frac{\partial}{\partial \xi} \right)^j h_2(x, \xi).$$

$$(1.1.9)$$

The right-hand side reduces to a finite sum and yields an exact formula, in the case when one is dealing with symbols of differential operators (of course, these are not Hilbert–Schmidt), since these are polynomial with respect to the second variable. That the formula is still correct in some suitable asymptotic sense, when h_1 and h_2 lie in appropriate (nonpolynomial) classes of symbols, can be proved to be true in a variety of cases, and is essential in applications of pseudodifferential analysis to partial differential equations.

The sharp composition of symbols reveals a quite different structure when emphasis is put on the first covariance formula (1.1.2). The algebra of differential operators on $\mathbb{R}^2$ which commute with the action of $SL(2, \mathbb{R})$ by linear changes of coordinates is generated by the single Euler operator

$$\mathcal{E} = (2i\pi)^{-1} \left(x \frac{\partial}{\partial x} + \xi \frac{\partial}{\partial \xi} + 1 \right), \quad (1.1.10)$$

the generalized eigenfunctions of which are just the functions on $\mathbb{R}^2 \setminus \{0\}$ homogeneous of degrees $-1 - i\lambda$, $\lambda \in \mathbb{R}$. Hence, there is an *exact* composition formula enabling one to expand $h_1 \# h_2$, under the assumption that h_1 (resp. h_2) is homogeneous of degree $-1 - i\lambda_1$ (resp. $-1 - i\lambda_2$), as the integral with respect to $\lambda \in \mathbb{R}$ of a family of functions homogeneous of degrees $-1 - i\lambda$. This formula can be found in [36, Theorem 17.1] and we shall not reproduce it here: let us only mention that it is more difficult to handle than any of the previous versions, since it involves the consideration of operators on the line with singular integral kernels. But the question is not which formula is more pleasant or easier to use: rather, what is the range of applicability of each.

For our present purpose, it is definitely the second (more complicated) formula which is important for comparison, and this will take us to our last topic discussed in connection with the Weyl calculus, to wit that of modular forms. Before coming to it, let us remind the reader that the so-called quasiregular representation of $SL(2,\mathbb{R})$ in the space $L^2_{\text{even}}(\mathbb{R}^2)$ is the one such that $(g.h)(x,\xi) = h(g^{-1}(x,\xi))$. It decomposes [10] as the direct integral of irreducible representations $\pi_{i\lambda}$, $\lambda \in \mathbb{R}$, the decomposition corresponding precisely to the decomposition of an even function $h \in L^2(\mathbb{R}^2)$ into its homogeneous components. Also, $\pi_{i\lambda}$ and $\pi_{-i\lambda}$ are unitarily equivalent, the intertwining operator being in this realization the symplectic Fourier transformation (the one with integral kernel $((x,\xi);(y,\eta)) \mapsto e^{2i\pi(x\eta - y\xi)}$, which commutes with all linear transformations of $\mathbb{R}^2$ associated with matrices in $SL(2,\mathbb{R})$, not only those in $SO(2)$). Each representation $\pi_{i\lambda}$ has another important realization in a Hilbert space of eigenfunctions, in the hyperbolic upper half-plane $\Pi \simeq SL(2,\mathbb{R})/SO(2)$, of the hyperbolic Laplacian Δ. Now, the Weyl calculus provides a simple characterization of even functions or tempered distributions on $\mathbb{R}^2$ by means of pairs of functions in Π, as follows: consider the first two (normalized) eigenfunctions u_i and u_i^1 of the harmonic oscillator $L = \pi \left(x^2 - \frac{1}{4\pi^2} \frac{d^2}{dx^2} \right)$ on the line: explicitly, $u_i(x) = 2^{\frac{1}{4}} e^{-\pi x^2}$ and $u_i^1(x) = 2^{\frac{5}{4}} \pi^{\frac{1}{2}} x e^{-\pi x^2}$. Given some $\tilde{g}$ lying above $g = \left(\begin{smallmatrix} a & b \\ c & d \end{smallmatrix} \right) \in SL(2,\mathbb{R})$, the image of u_i or u_i^1 under the transformation $\text{Met}(\tilde{g})$ only depends, up to the multiplication by some complex number of absolute value 1, on the point $z = \frac{ai+b}{ci+d}$. It is then possible, with some more or less arbitrary choice of the "phase factors," to define two families $(u_z)_{z\in\Pi}$ and $(u_z^1)_{z\in\Pi}$, coinciding with u_i and u_i^1 at $z = i$, each family being essentially permuted under any transformation in the image of the metaplectic representation. One can then characterize [36, p. 16] a symbol $\mathfrak{S} \in \mathcal{S}'_{\text{even}}(\mathbb{R}^2)$ by the pair (f_0, f_1) of functions on Π such that

$$f_0(z) = (u_z | \text{Op}(\mathfrak{S}) u_z),$$
$$f_1(z) = (u_z^1 | \text{Op}(\mathfrak{S}) u_z^1). \qquad (1.1.11)$$

Now, under each of these transfers, the quasiregular action of $SL(2,\mathbb{R})$ in $\mathcal{S}'_{\text{even}}(\mathbb{R}^2)$ transforms to the quasiregular action of this group on functions defined on Π: the difference is that, on Π, elements of G act by fractional-linear changes of the (complex) coordinate rather than linear ones. Besides, the operator $\pi^2 \mathcal{E}^2$ transforms to $\Delta - \frac{1}{4}$, so that, from [10] again, the transfers under study intertwine the two classical realizations of the principal series $(\pi_{i\lambda})$.

This is especially interesting in the automorphic situation, i.e., when the distributions $\mathfrak{S}$ under study are invariant under the linear changes of coordinates in $\mathbb{R}^2$ associated with matrices in a given arithmetic group, say $SL(2,\mathbb{Z})$: then, the pair (f_0, f_1) consists of automorphic functions and can be identified [36, p. 30] with a pair of Cauchy data for the Lax–Phillips scattering theory relative to the automorphic wave equation [18]. That pairs of automorphic functions have to be considered reflects the fact that the concept of automorphic distribution in $\mathbb{R}^2$ is slightly more precise than that of automorphic function in Π: for instance, the two nonholomorphic Eisenstein series $E_{\frac{1\pm i\lambda}{2}}$ (cf. any book on nonholomorphic modular form theory) are proportional, whereas the two *Eisenstein distributions* they come from are only

Fourier related. To put things in a different perspective, the decomposition of the quasiregular action of $SL(2, \mathbb{R})$ in $L^2_{\text{even}}(\mathbb{R}^2)$ involves each irreducible representation twice, since $\pi_{i\lambda} \simeq \pi_{-i\lambda}$. Dealing with automorphic distributions in the plane in place of pairs of automorphic functions has several advantages, one of which, discussed at length in [36], is that, after having solved a certain amount of technical difficulties – automorphic distributions are as a rule quite singular distributions – one may define and often compute the sharp product of two automorphic distributions as defined with the help of the Weyl calculus. Decomposing a sharp product such as $\mathfrak{S}_1 \# \mathfrak{S}_2$, where $\mathfrak{S}_j$ is an automorphic distribution homogeneous of degree $-1 - i\lambda_j$, as a superposition of automorphic distributions homogeneous of degree $-1 - i\lambda$, $\lambda \in \mathbb{R}$, requires that one should consider integral superpositions of Eisenstein distributions as well as series of so-called cusp distributions.

The metaplectic representation was introduced in full generality by Weil [41], but the "most interesting" case, that of a connected group containing the Fourier transformation, had already been used by physicists [28]. Many authors have discussed, in a spirit of harmonic analysis, some of the matters introduced so far, in particular the Heisenberg and metaplectic representations and their possible role in pseudodifferential analysis or modular form theory [5, 14], sometimes with a view toward developing new pseudodifferential techniques [31]. This is not our present point: on the contrary, we want to show that the metaplectic representation and Weyl calculus have a competitor, with just as many symmetries but not the same ones; we shall generally call this new point of view the *alternative* one, as opposed to the *usual* one as discussed so far.

We shall thus leave our discussion of the Weyl calculus at this point, but not before we have emphasized again one aspect of it, to wit the fact that the sharp composition of symbols brings to light bilinear operations $\mathcal{B}^{\lambda_1, \lambda_2}_\lambda$ (depending on $\lambda_1, \lambda_2, \lambda$) on nonholomorphic modular forms. These operations exist in two versions, depending on whether one realizes nonholomorphic modular forms, in the usual way, as functions on Π, or pairs of such as distributions in the plane. They were introduced and discussed in [35], then applied to the automorphic Weyl calculus in [36]. Of course, they also make sense in the nonautomorphic situation and their main property is expressed by the identity (also called a covariance relation)

$$\pi_{i\lambda}\left(\mathcal{B}^{\lambda_1, \lambda_2}_\lambda(h_1, h_2)\right) = \mathcal{B}^{\lambda_1, \lambda_2}_\lambda\left(\pi_{i\lambda_1}(h_1), \pi_{i\lambda_2}(h_2)\right). \tag{1.1.12}$$

This finally takes us to the central problem of the present work.

Instead of the principal series $(\pi_{i\lambda})_{\lambda \in \mathbb{R}}$ of representations of $SL(2, \mathbb{R})$, consider the holomorphic discrete series $(\mathcal{D}_{m+1})_{m=1,2,\dots}$, where the Hilbert space $\mathcal{H}_{m+1}$ of the representation consists of an appropriate space of holomorphic functions in Π (all details will be given in Sect. 2.1) and the representation is defined by the equation

$$\left(\mathcal{D}_{m+1}\left(\left(\begin{smallmatrix} a & b \\ c & d \end{smallmatrix}\right)\right) \chi\right)(z) = (-cz + a)^{-m-1} \chi\left(\frac{dz - b}{-cz + a}\right). \tag{1.1.13}$$

There is a family of bilinear operations $\mathcal{K}_{m+1}^{m_1+1,m_2+1}$, called the Rankin–Cohen brackets and defined by the equation

$$\mathcal{K}_{m+1}^{m_1+1,m_2+1}(\chi^1, \chi^2) = \sum_{q=0}^{p} (-1)^q \begin{pmatrix} m_1 + p \\ q \end{pmatrix} \begin{pmatrix} m_2 + p \\ p - q \end{pmatrix} (\chi^1)^{(p-q)} (\chi^2)^{(q)},$$

(1.1.14)

which satisfy just the covariance relation analogous to (1.1.12) obtained by replacing $\pi_{i\lambda}$ by $\mathcal{D}_{m+1}$ and making the other two required substitutions. In particular, they give rise to bilinear operations on holomorphic modular forms: these were introduced by Cohen [6] and, in some special instances involving one modular form only, had previously been introduced by Rankin [26].

Rankin–Cohen products also make their appearance, quite naturally, in connection with the theory of invariants. In [7], they played a role fitting more properly within star-product theory (cf. Remark 3.4.1 for a brief discussion of the relation between sharp products and this latter concept) than pseudodifferential analysis. That they occur, in some genuine quantization theory (or symbolic calculus of operators) as the various terms of the decomposition of sharp products of symbols, was seen in [34], and partly generalized to higher-rank cases in [8]. However, the phase space that had to be used in [34] was the one-sheeted hyperboloid, certainly a commendable homogeneous space, but one not quite as central as the phase space $\mathbb{R}^2$ of the Weyl calculus. One of the starting points of the present work was an attempt at making this substitution possible. As an impetus toward this, note the present ironic state of affairs: while it is the Rankin–Cohen brackets that are quite well known in modular form theory, it is the nonholomorphic operations $\mathcal{B}_\lambda^{\lambda_1,\lambda_2}$ that arise in a natural way from the universally known Weyl calculus. The latter ones are intrinsically more difficult to handle, which has certainly contributed to the fact that they have not yet succeeded in attracting the attention of number theorists.

Our first efforts to have Rankin–Cohen brackets appear as byproducts of the Weyl calculus proved doomed to fail, and the best one can do in this direction will be described in Sect. 5.2, especially Proposition 5.2.3. The first step toward the solution of our problem is easy: what we want is some representation of $SL(2,\mathbb{R})$ in $L^2(\mathbb{R}^2)$ which, in contrast to the quasiregular representation, will decompose not as a direct integral of representations from the principal series but as a direct sum of representations from the holomorphic discrete series. The simplest case of Howe's duality, to wit the fact that the pair $(SL(2,\mathbb{R}), SO(2))$ constitutes a dual reductive pair, calls for the consideration of the restriction Met$^{(2)}$ to $SL(2,\mathbb{R})$ of the metaplectic representation of the group Sp$(2, \mathbb{R})$ in $L^2(\mathbb{R}^2)$ (details are given in Sect. 2.1). Note that the representation Met$^{(2)}$, contrary to the quasiregular representation of the same group, does not act by changes of coordinates only since it involves on one hand the Euclidean Fourier transformation, on the other hand the operators of multiplication by functions $(x, \xi) \mapsto e^{i\pi c(x^2+\xi^2)}$ with $c \in \mathbb{R}$. The important point is that it decomposes as a discrete series of representations, containing (twice) all representations $\mathcal{D}_{m+1}$ with $m \geq 1$: the corresponding decomposition of elements of $L^2(\mathbb{R}^2)$ will be called the decomposition into isotypic components. This elementary case is just what is needed here: other cases of Howe's duality, and more sophisticated ways to make holomorphic discrete series or holomorphic modular forms appear from the decomposition of higher-dimensional metaplectic representations have been discussed by several authors [13, 9, 25, 20].

In alternative pseudodifferential analysis, spaces of symbols will be of a rather nonexotic nature, though $L^2(\mathbb{R}^2)$ would not do, and what takes the place of the quasiregular representation on the right-hand side of (1.1.2) is now known: it is the representation $\mathrm{Met}^{(2)}$. This is yet very far from solving our problem since we still have to define substitutes for the pair $(L^2(\mathbb{R}), \mathrm{Met}^{(1)})$ as well as for the Weyl calculus Op. Also, we want to preserve, in some modified form, the second covariance relation (1.1.3).

We would not have gone any further in this project if we had not, for some independent reason, constructed the *anaplectic* analysis. This construction [38], which had demanded a considerable amount of work in the higher-dimensional case, is much easier to describe in the one-dimensional case we are busying ourselves with in the present work. The simple idea imitates the fact that the one-dimensional metaplectic representation decomposes as the sum of two irreducible parts (restricting it to the even and odd parts of $L^2(\mathbb{R})$), which can be identified with the representations $\mathcal{D}_{\frac{1}{2}}$ and $\mathcal{D}_{\frac{3}{2}}$ from the discrete series of the twofold cover of $SL(2, \mathbb{R})$ (or its prolongation when the first one is concerned). In just the same way, but reversing the process, one can piece together two representations of $SL(2, \mathbb{R})$, one from the complementary series of this group, the other a (non unitarizable) signed version of the same, ending up with a representation of $SL(2, \mathbb{R})$ (the anaplectic representation) which still combines nicely with the Heisenberg representation. This new analysis contrasts violently with the usual one. First, the space which takes the place of $L^2(\mathbb{R})$ consists of functions which extend as entire functions and which usually increase wildly (not too wildly, though) at infinity. There is no useful scalar product, but there is an invariant nondegenerate pseudoscalar product instead (the same as a scalar product, except for positivity). A much more detailed presentation of anaplectic analysis has been given in loc.cit., and we shall satisfy ourselves (in Sect. 2.2) with a presentation of definitions and results.

Building the formula defining the alternative pseudodifferential calculus still proved a lengthy job: since the geometric ideas involved may be applicable to other situations, we shall briefly describe the way it was first obtained at the end of Sect. 3.1. However, the reader may dispense with this, provided he is ready to accept as it is the definition given in the beginning of the same section, which has some interest on its own right: it certainly emphasizes a fundamental difference between usual analysis and anaplectic analysis. In the first one, as is well known, the spectrum of the (already mentioned) harmonic oscillator L is the sequence $\frac{1}{2} + \mathbb{N}$. In anaplectic analysis, the spectrum of the formally identical operator is $\mathbb{Z}$ instead. A related fact is that the operator formally identical to the canonical annihilation operator $A = \pi^{\frac{1}{2}} (Q + iP)$, where Q and P are the standard position and momentum operators, respectively (Q is the multiplication by the variable x on the line and $P = (2i\pi)^{-1} \frac{d}{dx}$), becomes a linear automorphism of the basic space $\mathfrak{A}$ of functions on the line one has to consider. More generally, let us consider the operators

$A_z = \pi^{\frac{1}{2}} (Q - \bar{z}P)$, with z lying in the upper half-plane: up to scalar factors, they are pairwise conjugate under some (pseudo-unitary) transformations from the so-called anaplectic representation. The alternative pseudodifferential analysis is based on the use of integral superpositions of powers A_z^{-m-1}, $m = 1, 2, \ldots$. For comparison, let us recall that the usual pseudodifferential calculus uses instead exponentials of linear combinations, with pure imaginary coefficients, of Q and P.

Sections 3.1–3.4, devoted to the construction and analysis of the alternative pseudodifferential calculus, constitute the core of the present work. The more difficult part in Sect. 3.2 consists in giving a certain characterization, reminiscent of Beals' characterization [2] in usual pseudodifferential analysis, of appropriate classes of symbols by corresponding properties of the associated operators. One of the special features of the new symbolic calculus is that it decomposes into an ascending and a descending part: operators from the first one transform an eigenstate of any "harmonic oscillator" $L_z = A_z A_z^* - \frac{\mathrm{Im}\, z}{2}$ into the sum of a series of eigenstates of the same operator with higher energy level.

The characterization in Sect. 3.2 prepares the way for the study, in Sect. 3.4, of the sharp composition of symbols, by which is meant, as before, the bilinear operation on symbols corresponding to the composition of operators from the symbolic calculus under consideration. Coupling the sharp composition of symbols with their decomposition into isotypic components, one ends up with a family of bilinear operations $\mathcal{L}_{m+1}^{m_1+1,m_2+1}$, parametrized by a triple (m_1, m_2, m) of integers ≥ 1. Now, under some appropriate transferring map, the existence of which has already been alluded to, the restriction of the representation $\mathrm{Met}^{(2)}$ to the mth isotypic subspace of $L^2(\mathbb{R}^2)$ intertwines with the representation $\mathcal{D}_{m+1}$ from the discrete series of $SL(2, \mathbb{R})$ in the Hilbert space $\mathcal{H}_{m+1}$ of functions on the upper half-plane. Under a triple of such transfers, the family of bilinear operations $\mathcal{L}_{m+1}^{m_1+1,m_2+1}$ identifies, up to the multiplication by a simple constant, with the family of Rankin–Cohen products (or brackets) (1.1.14): this is not surprising since these operations are the only covariant bilinear differential operations compatible with the grading by the parameter specifying a representation from the discrete series of $SL(2, \mathbb{R})$ involved (or, in the arithmetic case, the weight of the holomorphic modular forms under consideration).

Putting in regard the decompositions of functions, or distributions, in the plane, into homogeneous or isotypic components, one may expect that the Euler operator on one hand, the rotation operator $\mathcal{R} = \xi \frac{\partial}{\partial x} - x \frac{\partial}{\partial \xi}$ on the other hand, should play reciprocal roles in the usual and alternative pseudodifferential analyses: this is true up to a surprising degree, as will be seen.

We hope to have convinced the reader, by now, that giving holomorphic modular forms a status comparable to that given to nonholomorphic forms justifies the introduction of alternative one-dimensional pseudodifferential analysis. We have not succeeded yet in finding an n-dimensional analogue: this does not imply that such a generalization does not exist, only that, if it does, it will require much work to bring it up. The reason for this has to do with anaplectic analysis itself: the higher-dimensional case, considered in [38], proved considerably more difficult to handle than the one-dimensional case. The main difficulty is that there is no class of simple functions, comparable to the Hermite functions of usual analysis, stable under the n-dimensional anaplectic representation. As long as we consider only the one-dimensional case, there is no question of applying alternative pseudodifferential analysis, or anaplectic analysis, to partial differential equations. However, this may not be the best way to approach the question: a more important one is the possible applicability of anaplectic analysis to Physics. We do not want to be too specific about this point: let us only mention (work in progress) that anaplectic analysis seems to be the right one to use when dealing with certain classically improperly posed special initial value problems, such as the ones which arise from relativistic mechanics when time and a space coordinate have been mixed up, in other words initial value problems as conceived by "tachyonic" observers; the structure of the pseudoscalar product of anaplectic analysis points toward the same direction.

In Chap. 4, we shall introduce a version depending on a real parameter ν mod 2 of anaplectic analysis. Then, the spectrum of the harmonic oscillator is the sequence $\nu + \frac{1}{2} + \mathbb{Z}$ and, at least when $\nu \notin \mathbb{Z}$, everything generalizes. The case discussed up to now corresponds to $\nu = -\frac{1}{2}$ and the case when $\nu \not\equiv 0$ mod 2 provides an extension of sorts of usual analysis. One last word: it was impossible to avoid the word "pseudodifferential" in the present work. However, the operators obtained are not at all generalizations of differential operators: Sect. 5.2 may shed some more light on this basic difference between alternative pseudodifferential analysis and the usual one.

Chapter 2
The Metaplectic and Anaplectic Representations

In this chapter, we briefly review some basic aspects of the metaplectic representation, especially in the one-dimensional and two-dimensional cases. Then, we shall introduce the new anaplectic analysis on the real line, in which the spectrum of the harmonic oscillator is $\mathbb{Z}$ rather than $\frac{1}{2} + \mathbb{N}$. The basic space $\mathfrak{A}$ substituting for $L^2(\mathbb{R})$ consists of functions on the line extending as entire functions, typically increasing like "bad" Gaussian functions at infinity. Nevertheless, there is on $\mathfrak{A}$ a well-defined translation-invariant concept of integral, and (in place of the scalar product of $L^2(\mathbb{R})$) a pseudoscalar product reminiscent of indefinite forms occurring in Physics. All symmetries of usual analysis expressing themselves by means of such objects as the Heisenberg representation, the Fourier transformation, and, more generally, the metaplectic representation, have counterparts in anaplectic analysis. Note that in Sect. 4.1, we shall have to consider the parameter-dependent ν-anaplectic analysis. The one considered in the present chapter (in Sect. 2.2) corresponds to $\nu = -\frac{1}{2}$: it will also be shown in Sect. 4.2 that the case when $\nu = 0$ yields an analysis containing the usual one.

2.1 The Metaplectic Representation

In this book, we are only interested in the case when the dimension n is 1 or 2: it will save space and add to the understanding, not to specify n from the start.

The symplectic group $\mathrm{Sp}(n, \mathbb{R})$ is the group of linear transformations g of $\mathbb{R}^n \times \mathbb{R}^n$, in block-form $g = \left(\begin{smallmatrix} A & B \\ C & D \end{smallmatrix} \right)$, which preserve the canonical symplectic form: this means that, if one sets $[(x, \xi), (y, \eta)] = -\langle x, \eta \rangle + \langle y, \xi \rangle$, the equation $[(x, \xi), (y, \eta)] = [g(x, \xi), g(y, \eta)]$ holds for any pair of points (x, ξ) and (y, η) in $\mathbb{R}^n \times \mathbb{R}^n$. In other words, one should have

$$ C'A = A'C, \quad D'B = B'D, \quad D'A - B'C = I, \tag{2.1.1} $$

the accent denoting the transposition map. The symplectic group is connected but its fundamental group is $\mathbb{Z}$: in particular, it has a (unique, up to isomorphism) twofold cover, called the metaplectic group, here denoted as $\widetilde{\mathrm{Sp}}(n,\mathbb{R})$: note that $\mathrm{Sp}(1,\mathbb{R}) = SL(2,\mathbb{R})$.

It is a fundamental fact [41] that there exists a unique unitary representation $\mathrm{Met}^{(n)}$ – the metaplectic representation – of $\widetilde{\mathrm{Sp}}(n,\mathbb{R})$ in $L^2(\mathbb{R}^n)$, satisfying the following properties:

(i) if C is a real symmetric $n \times n$-matrix, and if the identity $(2n) \times (2n)$-matrix is connected to the block-matrix $g = \begin{pmatrix} I & 0 \\ C & I \end{pmatrix}$ by means of the path $t \mapsto \begin{pmatrix} I & 0 \\ tC & I \end{pmatrix}$, finally if $\tilde{g}$ is the end of the path, in the metaplectic group, originating at the identity and covering the path within $\mathrm{Sp}(n,\mathbb{R})$ just defined, then the transformation $\mathrm{Met}^{(n)}(\tilde{g})$ is the multiplication by the function $x \mapsto \exp\left(i\pi\langle Cx, x \rangle\right)$;

(ii) if one considers the path, in the metaplectic group, originating at the identity and covering the path $t \mapsto \begin{pmatrix} (\cos t)I & (\sin t)I \\ (-\sin t)I & (\cos t)I \end{pmatrix}$ in the symplectic group, then the element $\tilde{g}$ reached for $t = \frac{\pi}{2}$ gives rise to the transformation $\mathrm{Met}^{(n)}(\tilde{g}) = e^{-\frac{i\pi n}{4}}\,\mathcal{F}$, where $\mathcal{F}$ is the usual Fourier transformation:

$$(\mathcal{F}u)(\xi) = \int_{\mathbb{R}^n} u(x)\,e^{-2i\pi\langle x,\xi\rangle}\,dx; \tag{2.1.2}$$

(iii) if $A \in GL^+(n,\mathbb{R})$ and $g = \begin{pmatrix} A & 0 \\ 0 & A'^{-1} \end{pmatrix}$, finally if $\tilde{g} \in \widetilde{\mathrm{Sp}}(n,\mathbb{R})$ is the end of a path originating at the identity of that group and covering a path $t \mapsto \begin{pmatrix} A_t & 0 \\ 0 & A_t'^{-1} \end{pmatrix}$ with $A_t \in GL(n,\mathbb{R})$ for all t, then $\mathrm{Met}^{(n)}(\tilde{g})$ is the transformation $u \mapsto u_1$, $u_1(x) = (\det A)^{-\frac{1}{2}} u(A^{-1}x)$.

The two metaplectic transformations associated with distinct points of $\widetilde{\mathrm{Sp}}(n,\mathbb{R})$ above the same point of $\mathrm{Sp}(n,\mathbb{R})$ differ only by the factor -1. The metaplectic representation is unitary in $L^2(\mathbb{R}^n)$; each transformation $\mathrm{Met}^{(n)}(\tilde{g})$ preserves the Schwartz space $\mathcal{S}(\mathbb{R}^n)$ and has a unique extension as a (weakly) continuous linear automorphism of $\mathcal{S}'(\mathbb{R}^n)$. The metaplectic representation is reducible: its irreducible subspaces are the two subspaces of $L^2(\mathbb{R}^n)$ characterized by parity.

To understand fully the metaplectic representation in a way not making it necessary to decompose symplectic matrices as products of the generators just defined, it is useful to characterize a function $u \in L^2(\mathbb{R}^n)$ or, more generally, a distribution in $\mathcal{S}'(\mathbb{R}^n)$, by its *quadratic transform*. Denote as $\mathrm{Sym}_n^{\mathbb{C}}$ the set of symmetric matrices with complex entries. The quadratic transform of a function $u \in L^2(\mathbb{R}^n)$ is the pair $((\mathcal{M}u)_0, (\mathcal{M}u)_1)$ of functions defined on the (Siegel) domain $(\mathrm{Sym}_n^{\mathbb{C}})_+$ consisting of all matrices $\sigma \in \mathrm{Sym}_n^{\mathbb{C}}$ with a positive definite real part, defined as follows:

$$(\mathcal{M}u)_0(\sigma) = \int_{\mathbb{R}^n} e^{-\pi\langle\sigma x, x\rangle}\,u(x)\,dx,$$

$$(\mathcal{M}u)_1(\sigma) = \int_{\mathbb{R}^n} (I+i\sigma)x \cdot e^{-\pi\langle\sigma x,x\rangle}\,u(x)\,dx. \tag{2.1.3}$$

Note that the function $(\mathcal{M}u)_1$ is vector valued: it is not necessary to bother with it if interested only in even functions u. The metaplectic representation $\mathrm{Met}^{(n)}$ of $\widetilde{\mathrm{Sp}}(n,\mathbb{R})$ in $L^2(\mathbb{R}^n)$ can be traced on the $\mathcal{M}$-transform as follows. For every element $\tilde{g}$ of the metaplectic group $\widetilde{\mathrm{Sp}}(n,\mathbb{R})$ above some element $g = \left(\begin{smallmatrix} A & B \\ C & D \end{smallmatrix}\right)$ of the symplectic group, there is a continuous choice of a determination of the square root of $\det(iB'\sigma + D')$ for $\sigma \in (\mathrm{Sym}_n^{\mathbb{C}})_+$ such that, for every $u \in L^2(\mathbb{R}^n)$, the following pair of equations holds:

$$(\mathcal{M}\,\mathrm{Met}^{(n)}(\tilde{g})\,u)_0(\sigma) = [\det(iB'\sigma + D')]^{-\frac{1}{2}}(\mathcal{M}u)_0((A'\sigma - iC')(iB'\sigma + D')^{-1}),$$

$$(\mathcal{M}\,\mathrm{Met}^{(n)}(\tilde{g})\,u)_1(\sigma) = [\det(iB'\sigma + D')]^{-\frac{1}{2}}(I + i\sigma)$$

$$\times [iB'\sigma + D' + i(A'\sigma - iC')]^{-1} \times (\mathcal{M}u)_1((A'\sigma - iC')(iB'\sigma + D')^{-1}).$$

$$(2.1.4)$$

Quadratic transforms will again, in Sects. 2.2 and 4.1, facilitate our understanding of the anaplectic and ν-anaplectic representations. This characterization, up to a sign $\pm$ depending only on $\tilde{g}$, not on σ, of the metaplectic transformation associated with $\tilde{g}$ can be found in [38, p. 100]. The sign can be obtained as soon as $\tilde{g}$ has been defined in full (i.e., as soon as the homotopy class of a path linking, in the group $\mathrm{Sp}(n,\mathbb{R})$, the identity to the element g above which $\tilde{g}$ is lying has been specified) by continuity.

We shall be especially interested in the subgroup of $\mathrm{Sp}(n,\mathbb{R})$, which is the image of $SL(2,\mathbb{R})$ under the embedding $\left(\begin{smallmatrix} a & b \\ c & d \end{smallmatrix}\right) \mapsto \left(\begin{smallmatrix} aI & bI \\ cI & dI \end{smallmatrix}\right)$. There is no harm in denoting this group simply as $SL(2,\mathbb{R})$, which we shall do from now on: when dealing with such a matrix, the superscript of the expression $\mathrm{Met}^{(n)}$ will make it clear whether we have in mind the image, under the metaplectic representation of the appropriate dimension, of the first 2×2-matrix or of the associated $(2n) \times (2n)$-matrix. As it turns out, if the dimension n is even, every loop within $SL(2,\mathbb{R})$ lifts as a loop in $\widetilde{\mathrm{Sp}}(n,\mathbb{R})$. Indeed, since $\mathbb{R}^n \sim \mathbb{R}^2 \otimes \mathbb{R}^{\frac{n}{2}}$, it entails no loss of generality to prove this only in the case when $n = 2$. Set $R_t = \left(\begin{smallmatrix} (\cos t)I & (-\sin t)I \\ (\sin t)I & (\cos t)I \end{smallmatrix}\right)$: when t moves on $[0, 2\pi]$, this is a loop in $\mathrm{Sp}(2,\mathbb{R})$, the image of a loop in $SL(2,\mathbb{R})$ generating the fundamental group of that space. Consider the two symplectic matrices

$$K_t = \begin{pmatrix} \cos t & 0 & -\sin t & 0 \\ 0 & 1 & 0 & 0 \\ \sin t & 0 & \cos t & 0 \\ 0 & 0 & 0 & 1 \end{pmatrix} \quad \text{and} \quad J = \begin{pmatrix} 0 & 1 & 0 & 0 \\ 1 & 0 & 0 & 0 \\ 0 & 0 & 0 & 1 \\ 0 & 0 & 1 & 0 \end{pmatrix}: \tag{2.1.5}$$

one may verify that $R_t = K_t J K_t J^{-1}$. Connecting J to the identity matrix within $\mathrm{Sp}(2,\mathbb{R})$, one sees that the loop R_t is equivalent to the loop $t \mapsto K_t^2 = K_{2t}$, which lifts as a loop in the twofold cover of that group. This implies that, when the dimension n is even, one can, for any $g \in SL(2,\mathbb{R})$, define $\mathrm{Met}^{(n)}(g)$ without any sign ambiguity, a fact which we shall take advantage of, presently, in the case when $n = 2$.

In this case, the definition of $\mathrm{Met}^{(2)}$ on generators of $SL(2,\mathbb{R})$ simplifies as follows (recall that $\left(\begin{smallmatrix} a & b \\ c & d \end{smallmatrix}\right)$ is to be identified with $\left(\begin{smallmatrix} aI & bI \\ cI & dI \end{smallmatrix}\right)$):

(i) $\left(\mathrm{Met}^{(2)}\left(\left(\begin{smallmatrix} 1 & 0 \\ c & 1 \end{smallmatrix}\right)\right)u\right)(x) = u(x)\,e^{i\pi c|x|^2}, \quad x \in \mathbb{R}^2;$

(ii) $\mathrm{Met}^{(2)}\left(\left(\begin{smallmatrix} 0 & 1 \\ -1 & 0 \end{smallmatrix}\right)\right)u = -i\,\mathcal{F}u;$

(iii) $\left(\mathrm{Met}^{(2)}\left(\left(\begin{smallmatrix} a & 0 \\ 0 & a^{-1} \end{smallmatrix}\right)\right)u\right)(x) = a^{-1}u(a^{-1}x), \quad x \in \mathbb{R}^2, \, a > 0.$ \qquad (2.1.6)

Recall the following formula, due to Hecke or Bochner [29]: if a function $u \in \mathcal{S}(\mathbb{R}^n)$ is the product of some "solid" spherical harmonic of degree m (i.e., a homogeneous polynomial on $\mathbb{R}^n$ of degree m, harmonic in the usual sense) by a radial function $U = U(r)$, the Fourier transform of u has the same property, with the same spherical harmonic, the function U being replaced by the function V defined by the equation

$$V(r) = 2\pi i^{-m} r^{\frac{2-n}{2}-m} \int_0^\infty U(t)\,t^{\frac{n}{2}+m}\,J_{\frac{n-2}{2}+m}(2\pi rt)\,dt. \qquad (2.1.7)$$

Given $m \in \mathbb{Z}$, we shall denote as $L_m^2(\mathbb{R}^2)$ the subspace of $L^2(\mathbb{R}^2)$ consisting of functions h – the change from u to h at this point, in the two-dimensional case, is deliberate, in view of future use – satisfying the equation (in which the matrix is of course to be identified with the corresponding linear automorphism of $\mathbb{R}^2$)

$$h \cap \begin{pmatrix} \cos\theta & -\sin\theta \\ \sin\theta & \cos\theta \end{pmatrix} - e^{-im\theta}\,h \qquad (2.1.8)$$

for every $\theta \in \mathbb{R}$ mod 2π: the spaces $L_m^2(\mathbb{R}^2)$ are called the isotypic subspaces of $L^2(\mathbb{R}^2)$. As indicated by (2.1.6), the Hilbert space decomposition $L^2(\mathbb{R}^2) = \bigoplus_{m\in\mathbb{Z}} L_m^2(\mathbb{R}^2)$ is preserved under the restriction of the two-dimensional metaplectic representation to the image of $SL(2,\mathbb{R})$ in $\widetilde{\mathrm{Sp}}(2,\mathbb{R})$. Variants of Proposition 2.1.1 have been known for a long time (cf. for instance [39]), even though the metaplectic representation had not yet been given a formal definition. As mentioned in the introduction, it also fits with the simplest case of Howe's duality.

Proposition 2.1.1. *If* $m = 1, 2, \ldots,$ *set* $c_m = (2\pi)^{\frac{m-1}{2}}\,((m-1)!)^{-\frac{1}{2}}$. *For any function* $h \in L_m^2(\mathbb{R}^2)$ *and* z *in the upper half-plane, set*

$$(\Theta_{\pm m}h)(z) = z^{-m-1}\int_{\mathbb{R}^2}(x_1 \pm ix_2)^m\,e^{-\frac{i\pi}{z}|x|^2}\,h(x)\,dx, \qquad \mathrm{Im}\,z > 0. \qquad (2.1.9)$$

The map $c_m\Theta_{\pm m}$ *is an isometry from the Hilbert space* $L_{\pm m}^2(\mathbb{R}^2)$ *onto the Hilbert space* $\mathcal{H}_{m+1}$ *consisting of all holomorphic functions* χ *in the upper half-plane* Π *satisfying the condition*

$$\|\chi\|_{m+1}^2 := \int_\Pi |\chi(z)|^2\,(\mathrm{Im}\,z)^{m+1}\,d\mu(z) < \infty : \qquad (2.1.10)$$

we have denoted as $d\mu$ *the usual invariant measure* $d\mu(x+iy) = y^{-2}dx\,dy$ *on* Π. *Moreover, denote as* $\mathcal{D}_{m+1}$ *the representation (taken from the so-called holomorphic discrete series) of* $SL(2,\mathbb{R})$ *in* $\mathcal{H}_{m+1}$ *defined by*

$$(\mathcal{D}_{m+1}\left(\left(\begin{smallmatrix} a & b \\ c & d \end{smallmatrix}\right)\right)\chi)(z) = (-cz+a)^{-m-1}\chi\left(\frac{dz-b}{-cz+a}\right). \tag{2.1.11}$$

Then, the operator $\Theta_{\pm m}$ intertwines the restriction to the space $L^2_{\pm m}(\mathbb{R}^2)$ of the representation $\mathrm{Met}^{(2)}$ of $SL(2,\mathbb{R})$ in $L^2(\mathbb{R}^2)$ with the representation $\mathcal{D}_{m+1}$.

For $m = 0$, the same conclusion holds provided one defines the space $\mathcal{H}_1$ as the Hardy space consisting of holomorphic functions χ such that $\sup_{y>0}\int_{-\infty}^{\infty}|\chi(x+iy)|^2\,dx < \infty$, and one takes $c_0 = (2\pi)^{-\frac{1}{2}}$.

Proof. Consider first the case when $m \geq 1$. In view of (2.1.8), one has if $h \in L^2_m(\mathbb{R}^2)$ the equation

$$(\Theta_m h)\left(-\frac{1}{z}\right) = 2\pi(-z)^{m+1}\int_0^{\infty} r^{m+1} e^{i\pi z r^2} h(r,0)\,dr. \tag{2.1.12}$$

One may then write, setting $z = s+it$,

$$\begin{aligned}
\|\Theta_m h\|^2_{m+1} &= \int_{\Pi}|(\Theta_m h)\left(-\frac{1}{z}\right)|^2 |z|^{-2m-2}(\mathrm{Im}\,z)^{m+1}\,d\mu(z) \\
&= 4\pi^2 \int_0^{\infty} t^{m-1}\,dt \int_{-\infty}^{\infty} ds \left|\int_0^{\infty} r^{m+1} e^{-\pi t r^2} e^{i\pi s r^2} h(r,0)\,dr\right|^2 \\
&= \pi^2 \int_0^{\infty} t^{m-1}\,dt \int_{\infty}^{\infty} ds \left|\int_0^{\infty} \rho^{\frac{m}{2}} e^{-\pi t \rho} e^{i\pi s \rho} h(\rho^{\frac{1}{2}},0)\,d\rho\right|^2 \\
&= 2\pi^2 \int_0^{\infty} t^{m-1}\,dt \int_0^{\infty} \rho^m e^{-2\pi t \rho}|h(\rho^{\frac{1}{2}},0)|^2\,d\rho \\
&= 2\pi^2 \times (2\pi)^{-m}(m-1)! \int_0^{\infty} |h(\rho^{\frac{1}{2}},0)|^2\,d\rho, \tag{2.1.13}
\end{aligned}$$

from which it is immediate to conclude that $c_m\Theta_m$ is an isometry. We may dispense with the proof that Θ_m is onto with the help of an irreducibility argument, after we have proved the intertwining properties (i), (ii), (iii). The first one is immediate, since $c - \frac{1}{z} = -[\frac{z}{1-cz}]^{-1}$ and $z^{-m-1}\times[\frac{z}{1-cz}]^{m+1} = (1-cz)^{-m-1}$; the third one is obtained after a change of variable. Starting from

$$[\Theta_m(\mathcal{F}h)](z) = z^{-m-1}\int_{\mathbb{R}^2} h(x)\,\mathcal{F}\left((x_1+ix_2)^m e^{-\frac{i\pi}{z}|x|^2}\right)dx, \tag{2.1.14}$$

one obtains the second one, namely

$$(\Theta_m(-i\mathcal{F}h))(z) = z^{-m-1}(\Theta_m h)\left(-\frac{1}{z}\right), \tag{2.1.15}$$

with the help of (2.1.7) and of the equation [21, p. 93]

$$\int_0^{\infty} J_m(2\pi|\xi|t)\,e^{-\frac{i\pi}{z}t^2} t^{m+1}\,dt = (2\pi|\xi|)^m\left(\frac{2i\pi}{z}\right)^{-m-1} e^{i\pi z|\xi|^2}. \tag{2.1.16}$$

The situation obtained when changing m to $-m$ can be reduced to the preceding one by means of the intertwining operator $h \mapsto h_1$, $h_1(x_1, x_2) = h(x_1, -x_2)$. Finally, when $m = 0$, only the norm computation has to be reconsidered. It follows the same lines (with a slight simplification), starting from the remark that

$$\| \Theta_0 h \|_1 = \| z \mapsto z^{-1} (\Theta_0 h) \left(-\frac{1}{z} \right) \|_1, \qquad (2.1.17)$$

a consequence of the unitarity of the representation $\mathcal{D}_1$. $\square$

Remark 2.1.1. The fact that the parameter m used in $L_m^2(\mathbb{R}^2)$ corresponds to the space $\mathcal{H}_{m+1}$ may often be felt as an inconvenience: however, there is nothing we can do about it.

We close this section with urging newcomers to pseudodifferential analysis to have another look at (1.1.1) and (1.1.2), now that their familiarity with the metaplectic representation may have been refreshed. Our aim in Chap. 3 (cf. introduction) is to introduce a new symbolic calculus, or "pseudodifferential analysis," for which a covariance formula somewhat similar to (1.1.2), but involving on the phase space $\mathbb{R}^2$ the representation $\mathrm{Met}^{(2)}$ in place of the quasiregular action $g \cdot \mathfrak{S} = \mathfrak{S} \circ g^{-1}$ of that group, would hold; at the same time, we want (1.1.3) to generalize too. The difficulty, as will be seen, is that everything has to be invented from scratch: as a space of possible functions u of one variable, we cannot use a space even remotely resembling $L^2(\mathbb{R})$; also, the one-dimensional metaplectic representation cannot play any role here. Then, Weyl's definition (1.1.1) has to be replaced by a new one.

The analysis to be developed in Sect. 2.2, rather than being regarded as an extension of the usual analysis, should be considered as alien to it. In Sect. 4.1, however, we shall imbed this analysis into a one-parameter ν-series: the case when $\nu = 0$ will then be shown to contain the part of usual analysis on the line centered around such objects as the Fourier transformation, the metaplectic representation, and Hermite functions.

2.2 Anaplectic Analysis

This section starts with a crash course on one-dimensional anaplectic analysis, a much more detailed version of which is to be found in [38]. Anaplectic analysis is just what is needed in the present context because we want to consider the inverse of the "annihilation" operator A (a name soon to be changed to that of "lowering" operator). In anaplectic analysis, the spectrum of the harmonic oscillator is $\mathbb{Z}$ rather than $\frac{1}{2} + \mathbb{N}$, and taking the inverse of A is all right, as will be recalled.

The basic difference between usual analysis and anaplectic analysis is the following. In the first one, there is a considerable supply (take for instance the Hermite functions) of functions on the line which extend as entire functions in the complex plane, while being simultaneously *very* rapidly decreasing at infinity. In anaplectic analysis, these two desirable properties have to be split between the function

u under consideration and other functions obtained from u, in a very specific way, with the help of the complex continuation process. This leads to the following definition, summed up, like a greater part of this short section, from the first two sections of [38].

Definition 2.2.1. Let us say that an entire function f of one variable is *nice* if on one hand $f(z)$ is bounded by a constant times some exponential $\exp(\pi R|z|^2)$, on the other hand the restriction of f to the *positive* half-line is bounded by a constant times some exponential $\exp(-\pi\varepsilon x^2)$: here, R and ε are assumed to be positive. The space $\mathfrak{A}$ consists of all entire functions u of one variable with the property that there exists a 4-tuple

$$\boldsymbol{f} = (f_0, f_1, f_{i,0}, f_{i,1}) \tag{2.2.1}$$

of nice functions such that

$$f_{i,0}(z) = \frac{1-i}{2}\left(f_0(iz) + i f_0(-iz)\right),$$
$$f_{i,1}(z) = \frac{1+i}{2}\left(f_1(iz) - i f_1(-iz)\right), \tag{2.2.2}$$

and such that the even part u_{even} of u coincides with the even part of f_0, and the odd part u_{odd} of u coincides with the odd part of f_1.

It can be proved, as a consequence of the Phragmén–Lindelöf lemma, that the vector-valued function $\boldsymbol{f}$ associated to $u \in \mathfrak{A}$ is necessarily unique. We shall call it the $\mathbb{C}^4$-realization of u. Here is a basic example.

Proposition 2.2.2. *Set, for x real,*

$$\phi(x) = (\pi|x|)^{\frac{1}{2}} I_{-\frac{1}{4}}(\pi x^2), \tag{2.2.3}$$

with [21, p. 66]

$$I_\nu(t) = \sum_{m \geq 0} \frac{\left(\frac{t}{2}\right)^{\nu+2m}}{m!\,\Gamma(\nu+m+1)} \tag{2.2.4}$$

for $t > 0$. The function ϕ lies in $\mathfrak{A}$. Its $\mathbb{C}^4$-realization is the function $\boldsymbol{f} = (\psi, 0, \psi, 0)$, with

$$\psi(x) = 2^{\frac{1}{2}}\,\pi^{-\frac{1}{2}}\,x^{\frac{1}{2}}\,K_{\frac{1}{4}}(\pi x^2) = (\pi x)^{\frac{1}{2}}\left[I_{-\frac{1}{4}}(\pi x^2) - I_{\frac{1}{4}}(\pi x^2)\right], \qquad x > 0. \tag{2.2.5}$$

The space $\mathfrak{A}$ is stable under the usual operators Q and P such that $(Qu)(x) = xu(x)$ and $(Pu)(x) = \frac{1}{2i\pi}u'(x)$. If the $\mathbb{C}^4$-realization $\boldsymbol{f}$ of u is the one given in (2.2.1), those of Qu and Pu are, respectively,

$$\boldsymbol{h}(z) = (z f_1(z), z f_0(z), z f_{i,1}(z), -z f_{i,0}(z)) \tag{2.2.6}$$

and

$$\boldsymbol{h} = \frac{1}{2i\pi}(f_1', f_0', -f_{i,1}', f_{i,0}'). \tag{2.2.7}$$

One may introduce, in the usual way, the harmonic oscillator and the operators

$$A^* = \pi^{\frac{1}{2}} \left(x - \frac{1}{2\pi} \frac{d}{dx} \right), \qquad A = \pi^{\frac{1}{2}} \left(x + \frac{1}{2\pi} \frac{d}{dx} \right). \qquad (2.2.8)$$

In usual analysis, these two operators would be called the creation and annihilation operators: however, for reasons to be seen immediately, they are to be called, now, the *raising* and *lowering* operators instead.

Theorem 2.2.3. *The spectrum of the harmonic oscillator*

$$L = \pi \left(Q^2 + P^2 \right) \qquad (2.2.9)$$

in the space $\mathfrak{A}$ *is* $\mathbb{Z}$, *and for every* $j \in \mathbb{Z}$ *the eigenspace corresponding to the eigenvalue* j *is generated by the function* ϕ^j, *with*

$$\phi^j = A^{*j} \phi \quad \text{if } j \geq 0, \qquad \phi^j = A^{|j|} \phi \quad \text{if } j \leq 0. \qquad (2.2.10)$$

There is on the space $\mathfrak{A}$ a useful nondegenerate *pseudoscalar* product $(\ |\ )$ (this is the same as a scalar product, except for positivity) defined in terms of the $\mathbb{C}^4$-realizations of the two functions involved as

$$(h \,|\, f) =$$
$$2^{\frac{1}{2}} \int_0^\infty \left(\bar{h}_0(x) f_0(x) + \bar{h}_1(x) f_1(x) + \bar{h}_{i,0}(x) f_{i,0}(x) - \bar{h}_{i,1}(x) f_{i,1}(x) \right) dx. \qquad (2.2.11)$$

The operators Q and P are self-adjoint on $\mathfrak{A}$ with respect to this pseudoscalar product. The functions ϕ^j, $j \in \mathbb{Z}$, are pairwise orthogonal with respect to it. The function ϕ is normalized and one has $(\phi^{k+1} \,|\, \phi^{k+1}) = (k + \frac{1}{2})(\phi^k \,|\, \phi^k)$ and $(\phi^{-k} \,|\, \phi^{-k}) = (-1)^k (\phi^k \,|\, \phi^k)$ for $k \geq 0$. Consequently,

$$(\phi^k \,|\, \phi^k) = \begin{cases} 2^{-2k} \frac{(2k)!}{k!}, & k \geq 0, \\ (-1)^k 2^{-2|k|} \frac{(2|k|)!}{|k|!}, & k < 0. \end{cases} \qquad (2.2.12)$$

In anaplectic analysis, the Heisenberg representation, as defined in a way formally identical to the usual one, preserves the anaplectic space $\mathfrak{A}$.

Theorem 2.2.4. *Given* $u \in \mathfrak{A}$ *and* $(y, \eta) \in \mathbb{C}^2$, *the function* $\exp(2i\pi(\eta Q - yP)) u$ *such that*

$$(\exp(2i\pi(\eta Q - yP)) u)(x) = u(x - y) e^{2i\pi(x - \frac{y}{2})\eta} \qquad (2.2.13)$$

lies in $\mathfrak{A}$ *too. If one restricts* (y, η) *to the space* $\mathbb{R}^2$, *the representation of Heisenberg's group (or, in an equivalent way, the projective representation of* $\mathbb{R}^2$) *so defined preserves the pseudoscalar product.*

Of course, a function such as ϕ, an eigenfunction of the harmonic oscillator for a classically forbidden eigenvalue, cannot be integrable on the real line. However, there is on $\mathfrak{A}$ a substitute for the notion of integral, still a translation-invariant linear form.

Proposition 2.2.5. *If $f = (f_0, f_1, f_{i,0}, f_{i,1})$ is the $\mathbb{C}^4$-realization of some function $u \in \mathfrak{A}$, set*

$$\text{Int}\,[u] = 2^{\frac{1}{2}} \int_0^\infty (f_0(x) + f_{i,0}(x))\,dx. \tag{2.2.14}$$

For every $y \in \mathbb{C}$, with $(e^{-2i\pi yP} u)(z) = u(z-y)$, one has

$$\text{Int}\,[e^{-2i\pi yP} u] = \text{Int}\,[u]. \tag{2.2.15}$$

This concept of integral makes the definition of an anaplectic Fourier transformation possible.

Proposition 2.2.6. *Given $x \in \mathbb{R}$, define the function e_x as $e_x(y) = e^{-2i\pi xy}$. For any $u \in \mathfrak{A}$, the anaplectic Fourier transform $\mathcal{F}_{\text{ana}}\, u$ of u defined as*

$$(\mathcal{F}_{\text{ana}}\, u)(x) = \text{Int}\,[e_x u] \tag{2.2.16}$$

lies in $\mathfrak{A}$ too. A fully developed version of the preceding definition, in terms of the $\mathbb{C}^4$-realization of u, is

$$(\mathcal{F}_{\text{ana}}\, u)(x) = 2^{\frac{1}{2}} \int_0^\infty f_0(y)\cos 2\pi xy\, dy - 2^{\frac{1}{2}} i \int_0^\infty f_1(y)\sin 2\pi xy\, dy$$

$$+ 2^{\frac{1}{2}} \int_0^\infty f_{i,0}(y)\cosh 2\pi xy\, dy - 2^{\frac{1}{2}} i \int_0^\infty f_{i,1}(y)\sinh 2\pi xy\, dy. \tag{2.2.17}$$

The function ϕ introduced in (2.2.3) is invariant under $\mathcal{F}_{\text{ana}}$.

It is essential to recall here the definition of the anaplectic representation (the substitute, in anaplectic analysis, of the metaplectic representation).

Theorem 2.2.7. *There is a unique representation* Ana *of $SL(2, \mathbb{R})$ in the space $\mathfrak{A}$ with the following properties:*

(i) if $g = \left(\begin{smallmatrix} 1 & 0 \\ c & 1 \end{smallmatrix}\right)$, one has $(\text{Ana}(g)\, u)(x) = u(x)\, e^{i\pi cx^2}$;

(ii) if $g = \left(\begin{smallmatrix} a & 0 \\ 0 & a^{-1} \end{smallmatrix}\right)$ with $a > 0$, one has $(\text{Ana}(g)\, u)(x) = a^{-\frac{1}{2}} u(a^{-1}x)$;

(iii) one has $\text{Ana}\left(\left(\begin{smallmatrix} 0 & 1 \\ -1 & 0 \end{smallmatrix}\right)\right) = \mathcal{F}_{\text{ana}}$.

This representation is pseudo-unitary, i.e., it preserves the scalar product introduced in (2.2.11). It combines with the (anaplectic) Heisenberg representation in the way characterized by the equation

$$\text{Ana}(g)\, e^{2i\pi(\eta Q - yP)}\, \text{Ana}(g^{-1}) = e^{2i\pi(\eta' Q - y'P)} \tag{2.2.18}$$

if $g = \left(\begin{smallmatrix} a & b \\ c & d \end{smallmatrix}\right) \in SL(2, \mathbb{R})$ and $g\left(\begin{smallmatrix} y \\ \eta \end{smallmatrix}\right) = \left(\begin{smallmatrix} y' \\ \eta' \end{smallmatrix}\right)$.

There is an extra benefit in anaplectic analysis: one can extend the anaplectic representation to that of the subgroup $SL_i(2, \mathbb{R})$ of $SL(2, \mathbb{C})$ generated by $SL(2, \mathbb{R})$ and by the matrix $g = \left(\begin{smallmatrix} -i & 0 \\ 0 & i \end{smallmatrix}\right)$, defining $\text{Ana}(g)$, in this case, as the transformation (which preserves $\mathfrak{A}$) $u \mapsto u_i$, $u_i(x) = u(ix)$. However, pseudo-unitarity is then lost.

Note that (in contradiction to the case of the one-dimensional metaplectic representation) one has a genuine representation of $SL(2, \mathbb{R})$, without it being necessary to use a twofold cover.

The infinitesimal version of (2.2.18) is the following: if $(y, \eta) \in \mathbb{C}^2$ and if $g = \left(\begin{smallmatrix} a & b \\ c & d \end{smallmatrix} \right) \in SL(2, \mathbb{R})$, one has

$$\text{Ana}(g) \, (\eta \, Q - y P) \, \text{Ana}(g^{-1}) = \eta' Q - y' P \qquad (2.2.19)$$

with $\left(\begin{smallmatrix} y' \\ \eta' \end{smallmatrix} \right) = \left(\begin{smallmatrix} a & b \\ c & d \end{smallmatrix} \right) \left(\begin{smallmatrix} y \\ \eta \end{smallmatrix} \right)$. It is used in the elementary proof of the proposition that follows.

Proposition 2.2.8. *Given $z \in \Pi$, the hyperbolic (Poincaré) upper half-plane, set*

$$A_z = \pi^{\frac{1}{2}} \, (Q - \bar{z} P), \qquad A_z^* = A_{\bar{z}} = \pi^{\frac{1}{2}} \, (Q - z P), \qquad (2.2.20)$$

and define

$$L_z = A_z A_z^* - \frac{\operatorname{Im} z}{2} = A_z^* A_z + \frac{\operatorname{Im} z}{2}. \qquad (2.2.21)$$

One then has the identities

$$\text{Ana}\left(\left(\begin{smallmatrix} a & b \\ c & d \end{smallmatrix} \right) \right) A_z \, \text{Ana}\left(\left(\begin{smallmatrix} d & -b \\ -c & a \end{smallmatrix} \right) \right) = (c\bar{z} + d) A_{\frac{az+b}{cz+d}},$$

$$\text{Ana}\left(\left(\begin{smallmatrix} a & b \\ c & d \end{smallmatrix} \right) \right) L_z \, \text{Ana}\left(\left(\begin{smallmatrix} d & -b \\ -c & a \end{smallmatrix} \right) \right) = |c\bar{z} + d|^2 L_{\frac{az+b}{cz+d}}. \qquad (2.2.22)$$

If one takes in particular $g_z = \begin{pmatrix} y^{\frac{1}{2}} & y^{-\frac{1}{2}} x \\ 0 & y^{-\frac{1}{2}} \end{pmatrix}$ if $z = x + iy$, the function

$$\phi_z^j = \text{Ana}(g_z) \, \phi^j, \qquad (2.2.23)$$

with ϕ^j as defined in Theorem 2.2.3, is a basis of the (one-dimensional) eigenspace of L_z in $\mathfrak{A}$ corresponding to the eigenvalue $j \operatorname{Im} z$.

In anaplectic analysis, it is often necessary to go back to the $\mathbb{C}^4$-realizations of functions. We shall have to use, later, the following, the proof of which is immediate: if $g = \left(\begin{smallmatrix} a & 0 \\ 0 & a^{-1} \end{smallmatrix} \right)$ with $a > 0$, $\text{Ana}(g)$ acts on $\mathbb{C}^4$-realizations as $f \mapsto h$, with

$$h(x) = (a^{-\frac{1}{2}} f_0(a^{-1}x), a^{-\frac{1}{2}} f_1(a^{-1}x), a^{-\frac{1}{2}} f_{i,0}(a^{-1}x), a^{-\frac{1}{2}} f_{i,1}(a^{-1}x)); \quad (2.2.24)$$

if $g = \left(\begin{smallmatrix} 1 & 0 \\ c & 1 \end{smallmatrix} \right)$, $\text{Ana}(g)$ acts as $f \mapsto h$, with

$$h(x) = (f_0(x) \, e^{i\pi c x^2}, f_1(x) \, e^{i\pi c x^2}, f_{i,0}(x) \, e^{-i\pi c x^2}, f_{i,1}(x) \, e^{-i\pi c x^2}). \qquad (2.2.25)$$

It is also necessary to examine the infinitesimal operators of the anaplectic representation, defined by the formula

$$d\text{Ana}(X) = \frac{1}{2i\pi} \frac{d}{dt} \Big|_{t=0} \text{Ana} \, (\exp tX), \qquad X \in \mathfrak{sl}(2, \mathbb{R}). \qquad (2.2.26)$$

Taking first $\exp tX = \begin{pmatrix} 1 & 0 \\ t & 1 \end{pmatrix}$, next $\exp tX = \begin{pmatrix} e^t & 0 \\ 0 & e^{-t} \end{pmatrix}$, one obtains from the cases (i) and (ii) of Theorem 2.2.7 the relations

$$d\text{Ana}\left(\begin{pmatrix} 0 & 0 \\ 1 & 0 \end{pmatrix}\right) = \frac{1}{2}Q^2,$$

$$d\text{Ana}\left(\begin{pmatrix} 1 & 0 \\ 0 & -1 \end{pmatrix}\right) = -\frac{1}{2i\pi}\left(x\frac{d}{dx} + \frac{1}{2}\right) = -\frac{1}{2}(QP + PQ). \tag{2.2.27}$$

Taking the conjugate of the first relation under the anaplectic Fourier transformation, one finds

$$d\text{Ana}\left(\begin{pmatrix} 0 & -1 \\ 0 & 0 \end{pmatrix}\right) = \frac{1}{2}P^2. \tag{2.2.28}$$

As a consequence,

$$d\text{Ana}\left(\begin{pmatrix} 0 & 1 \\ -1 & 0 \end{pmatrix}\right) = -\frac{1}{2}(Q^2 + P^2) \tag{2.2.29}$$

so that

$$\exp(-itL) = \text{Ana}\left(\begin{pmatrix} \cos t & \sin t \\ -\sin t & \cos t \end{pmatrix}\right). \tag{2.2.30}$$

Of course, no Stone's theorem is available in $\mathfrak{A}$, a space with a pseudoscalar product only: what is meant by the equation that precedes is that, if one defines $\exp(-itL)$ by this equation, one obtains a one-parameter group of operators satisfying the right differential equation. Since $L\phi = 0$, this equation proves the invariance of ϕ under the operator in (2.2.30). One may note the equation $\mathcal{F}_{\text{ana}} = \exp(-\frac{i\pi}{2}L)$: in usual analysis, there is an extra factor $e^{-\frac{i\pi}{4}}$ on the left-hand side, of course linked to the shift by $\frac{1}{2}$ of the spectrum of the harmonic oscillator.

A useful corollary of (2.2.30), together with Theorem 2.2.3, is the following generalization of (2.2.23): if $z \in \Pi$, $g = \begin{pmatrix} a & b \\ c & d \end{pmatrix}$ and $j \in \mathbb{Z}$, one has

$$\text{Ana}(g)\,\phi_z^j = \left(\frac{cz+d}{|cz+d|}\right)^j \phi_{\frac{az+b}{cz+d}}^j. \tag{2.2.31}$$

The proof in the case when $z = i$ goes as follows: assuming that $z = \frac{ai+b}{ci+d}$, write

$$\begin{pmatrix} a & b \\ c & d \end{pmatrix} = \begin{pmatrix} y^{\frac{1}{2}} & y^{-\frac{1}{2}}x \\ 0 & y^{-\frac{1}{2}} \end{pmatrix}\begin{pmatrix} \frac{d}{(c^2+d^2)^{\frac{1}{2}}} & -\frac{c}{(c^2+d^2)^{\frac{1}{2}}} \\ \frac{c}{(c^2+d^2)^{\frac{1}{2}}} & \frac{d}{(c^2+d^2)^{\frac{1}{2}}} \end{pmatrix}$$

$$= \begin{pmatrix} y^{\frac{1}{2}} & y^{-\frac{1}{2}}x \\ 0 & y^{-\frac{1}{2}} \end{pmatrix}\begin{pmatrix} \cos t & \sin t \\ -\sin t & \cos t \end{pmatrix}, \tag{2.2.32}$$

finding, as a result of (2.2.30),

$$\text{Ana}(g)\,\phi^j = e^{-ijt}\,\text{Ana}(g_z)\,\phi^j = \left(\frac{ci+d}{|ci+d|}\right)^j \phi_z^j. \tag{2.2.33}$$

The general case then follows from the equation $\begin{pmatrix} a & b \\ c & d \end{pmatrix} \begin{pmatrix} y^{\frac{1}{2}} & y^{-\frac{1}{2}}x \\ 0 & y^{-\frac{1}{2}} \end{pmatrix} = \begin{pmatrix} a_1 & b_1 \\ c_1 & d_1 \end{pmatrix}$ with $c_1 i + d_1 = y^{-\frac{1}{2}}(cz+d)$.

Lemma 2.2.9 will be necessary in Sects. 3.1 and 3.2.

Lemma 2.2.9. *One has*

$$A_z \phi_z^k = \gamma_k (\operatorname{Im} z)^{\frac{1}{2}} \phi_z^{k-1}, \qquad A_z^* \phi_z^k = \gamma_k^* (\operatorname{Im} z)^{\frac{1}{2}} \phi_z^{k+1}, \tag{2.2.34}$$

with

$$\gamma_k = \begin{cases} k - \frac{1}{2} & \text{if } k \geq 1, \\ 1 & \text{if } k \leq 0 \end{cases}, \qquad \gamma_k^* = \begin{cases} 1 & \text{if } k \geq 0, \\ k + \frac{1}{2} & \text{if } k \leq -1 \end{cases}. \tag{2.2.35}$$

One has the relations

$$\left[4i (\operatorname{Im} z) \frac{\partial}{\partial z} - j \right] \phi_z^j = -\gamma_j \gamma_{j-1} \phi_z^{j-2},$$

$$\left[4i (\operatorname{Im} z) \frac{\partial}{\partial \bar{z}} - j \right] \phi_z^j = -\gamma_j^* \gamma_{j+1}^* \phi_z^{j+2}. \tag{2.2.36}$$

Proof. Relations (2.2.34) are a consequence of (2.2.21) and (2.2.22) together with the fact, also indicated in Proposition 2.2.8, that ϕ_z^k is an eigenfunction of L_z corresponding to the eigenvalue $k \operatorname{Im} z$.

With $J = \begin{pmatrix} 0 & 1 \\ -1 & 0 \end{pmatrix}$ and $z = x + iy$, set, recalling that the matrix g_z has been introduced in Proposition 2.2.8,

$$\tilde{g}_z : = J g_z J^{-1} = \begin{pmatrix} 1 & 0 \\ -x & 1 \end{pmatrix} \begin{pmatrix} y^{-\frac{1}{2}} & 0 \\ 0 & y^{\frac{1}{2}} \end{pmatrix} : \tag{2.2.37}$$

it follows that, for every $u \in \mathfrak{A}$,

$$(\operatorname{Ana}(\tilde{g}_z) u)(t) = e^{-i\pi x t^2} y^{\frac{1}{4}} u(y^{\frac{1}{2}} t). \tag{2.2.38}$$

On the other hand, from (2.2.30) and Theorem 2.2.3, we obtain

$$\operatorname{Ana}(J) \phi^j = (-i)^j \phi^j, \tag{2.2.39}$$

so that

$$\phi_z^j = (-i)^j \operatorname{Ana}(J^{-1}) \operatorname{Ana}(\tilde{g}_z) \phi^j. \tag{2.2.40}$$

Next, using the fact that $\pi (Q^2 + P^2) \phi^j = j \phi^j$ and Heisenberg's relation $[P, Q] = \frac{1}{2i\pi}$, one obtains

$$A^2 \phi^j = 2\pi Q (Q + iP) \phi^j + (\frac{1}{2} - j) \phi^j,$$

$$A^{*2} \phi^j = 2\pi Q (Q - iP) \phi^j - (\frac{1}{2} + j) \phi^j. \tag{2.2.41}$$

On the other hand, a direct computation, starting from (2.2.38), shows that

$$\left[\mathrm{Ana}(\tilde{g}_z)\left(Q^2 u\right)\right](t) = -\frac{1}{i\pi}\, y\, \frac{\partial}{\partial x}\left(\mathrm{Ana}(\tilde{g}_z)\, u\right)(t),$$

$$\left[\mathrm{Ana}(\tilde{g}_z)\left(QP u\right)\right](t) = \frac{1}{i\pi}\left(y\frac{\partial}{\partial y} - \frac{1}{4}\right)\left(\mathrm{Ana}(\tilde{g}_z)\, u\right)(t) \qquad (2.2.42)$$

for every function $u \in \mathfrak{A}$. It follows that

$$\mathrm{Ana}(\tilde{g}_z) A^2 \, \phi^j = \left(4iy\frac{\partial}{\partial z} - j\right) \mathrm{Ana}(\tilde{g}_z)\, \phi^j,$$

$$\mathrm{Ana}(\tilde{g}_z) A^{*2} \, \phi^j = \left(4iy\frac{\partial}{\partial \bar{z}} - j\right) \mathrm{Ana}(\tilde{g}_z)\, \phi^j. \qquad (2.2.43)$$

Equations (2.2.36) follow if one uses (2.2.40) and (2.2.34). $\qquad\qquad\qquad\square$

We need to introduce some Hilbert space methods in anaplectic analysis. True, (2.2.11) only introduces a pseudoscalar product on $\mathfrak{A}$: however, its restriction to the even part of $\mathfrak{A}$ is positive definite while, on the odd part, one may take advantage of the linear isomorphism provided by an operator changing the parity, for instance the canonical lowering operator. The following is taken from [38, p. 154].

Proposition 2.2.10. *Let* $(\phi_j)_{j \in \mathbb{Z}}$ *be the sequence of eigenfunctions of the anaplectic harmonic oscillator introduced in Theorem 2.2.3. Given any function* $u \in \mathfrak{A}$, *the set of scalar products of* u *against the functions* ϕ_j *satisfies for some constants* $C > 0$ *and* $\delta \in]0,1[$ *the estimate*

$$|(\phi^j \,|\, u)| \leq C\,[\frac{|j|}{2}]!\,(2\delta)^{\frac{|j|}{2}}, \qquad j \in \mathbb{Z}. \qquad (2.2.44)$$

Conversely, given any sequence $(a_j)_{j \in \mathbb{Z}}$ *of complex numbers satisfying for some* $C > 0$ *and* $\delta \in]0,1[$ *the inequality*

$$|a_j| \leq C\,[\frac{|j|}{2}]!\,(2\delta)^{\frac{|j|}{2}}, \qquad j \in \mathbb{Z}, \qquad (2.2.45)$$

there exists a unique function $u \in \mathfrak{A}$ *such that* $a_j = (\phi^j \,|\, u)$ *for all* j.

Recalling (2.2.12), one sees that the orthogonal set $(\psi_j)_{j \in 2\mathbb{Z}}$, with

$$\psi^j = 2^{|j|}\left(\frac{|j|!}{(2|j|)!}\right)^{\frac{1}{2}} \phi^j, \qquad (2.2.46)$$

consists of normalized functions. It was actually shown in loc.cit. that this set constitutes a Hilbert basis of the completion of $\mathfrak{A}_{\mathrm{even}}$ under the norm associated with the restriction to this space of the scalar product (2.2.11). Hence, if $u \in \mathfrak{A}_{\mathrm{even}}$, one has the expansion, convergent in the Hilbert sense,

$$u = \sum_{j \in 2\mathbb{Z}} \frac{2^{2|j|}|j|!}{(2|j|)!}\,(\phi^j \,|\, u)\,\phi^j : \qquad (2.2.47)$$

one can of course verify that the condition (2.2.44) makes the series $\sum |(\psi^j|u)|^2$ convergent.

When $u \in \mathfrak{A}$ is odd, we can write instead

$$u = \sum_{j \in 2\mathbb{Z}} \frac{2^{2|j|}|j|!}{(2|j|)!}\, (\phi^j|Au)A^{-1}\phi^j. \tag{2.2.48}$$

With the help of (2.2.34) and of the equation

$$A^* \phi^j = \phi^{j+1} \text{ if } j \geq 0, \qquad A^* \phi^j = (j+\tfrac{1}{2})\phi^{j+1} \text{ if } j < 0, \tag{2.2.49}$$

this can be written as

$$u = \sum_{j \text{ even} \geq 0} \frac{2^{2j} j!}{(2j)!}\, (j+\tfrac{1}{2})^{-1}\, (\phi^{j+1}|u)\,\phi^{j+1}$$

$$+ \sum_{j \text{ even} < 0} \frac{2^{-2j}(-j)!}{(-2j)!}\, (j+\tfrac{1}{2})\, (\phi^{j+1}|u)\,\phi^{j+1}. \tag{2.2.50}$$

The general formula, whether $u \in \mathfrak{A}$ has any definite parity or not, is thus

$$u - \sum_{\ell \in \mathbb{Z}} o_\ell\, (\phi^\ell|u)\,\phi^\ell, \tag{2.2.51}$$

with

$$c_\ell = \begin{cases} \dfrac{2^{2|\ell|}|\ell|!}{(2|\ell|)!} & \text{if } \ell \text{ is even,} \\[2ex] \dfrac{2^{2|\ell-1|}(|\ell-1|)!}{(2|\ell-1|)!}\,(\ell-\tfrac{1}{2})^{-\operatorname{sign}\ell} & \text{if } \ell \text{ is odd.} \end{cases} \tag{2.2.52}$$

Note that, if $\ell \geq 1$, one has $c_\ell = \frac{2^{2\ell} \ell!}{(2\ell)!}$ in both cases. Since the anaplectic representation is pseudo-unitary, one can also write

$$u = \sum_{\ell \in \mathbb{Z}} c_\ell\, (\phi^\ell_z|u)\,\phi^\ell_z \tag{2.2.53}$$

for any $z \in \Pi$, as a consequence of (2.2.23).

We end this section with another useful characterization, taken from [38, p. 7, 188–190] of the space $\mathfrak{A}$. The introduction of the quadratic transform $((\mathcal{Q}u)_0, (\mathcal{Q}u)_1)$ of u will be found more natural if compared with (2.1.3).

Proposition 2.2.11. *Let u be an entire function of one variable satisfying for some pair of constants C, R the estimate $|f(z)| \leq C\,e^{\pi R|z|^2}$. Set, for σ real and large,*

$$(\mathcal{Q}u)_0(\sigma) = \int_{-\infty}^{\infty} e^{-\pi\sigma x^2}\, u(xe^{-\frac{i\pi}{4}})\,dx\,,$$

$$(\mathcal{Q}u)_1(\sigma) = \int_{-\infty}^{\infty} (1+i\sigma)x\,e^{-\pi\sigma x^2}\, u(xe^{-\frac{i\pi}{4}})\,dx\,, \tag{2.2.54}$$

and, for z on the unit circle, $z = e^{-i\theta}$ with $\theta > 0$ and small,

$$(\mathcal{K}u)_0(z) = |1 - z|^{-\frac{1}{2}} (\mathcal{Q}u)_0 \left(i \frac{1+z}{1-z} \right), \qquad (2.2.55)$$

finally introducing a function $(\mathcal{K}u)_1$ linked to $(\mathcal{Q}u)_1$ by the same transformation as the one giving $(\mathcal{K}u)_0$ in terms of $(\mathcal{Q}u)_0$. The following three conditions are equivalent:

   (i) u lies in the space $\mathfrak{A}$;
   (ii) each of the two functions $(\mathcal{Q}u)_0$ and $(\mathcal{Q}u)_1$ extends as an analytic function on the real line, admitting for large $|\sigma|$ a convergent expansion $(\mathcal{Q}u)_j(\sigma) = \sum_{n \geq 0} a_n^{(j)} \sigma^{-n} |\sigma|^{-\frac{1}{2}}$;
   (iii) each of the two functions $(\mathcal{K}u)_0$ and $(\mathcal{K}u)_1$, initially defined in a neighborhood of the point $z = 1$ of the unit circle, extends as an analytic function to the full circle.

The $\mathcal{Q}$-realization of $\mathfrak{A}$ is especially useful when dealing with certain representation-theoretic aspects. Set

$$|\sigma|_0^{-1-\rho} = |\sigma|^{-1-\rho}, \qquad |\sigma|_1^{-1-\rho} = |\sigma|^{-1-\rho} \times \operatorname{sign}\sigma, \qquad \sigma \in \mathbb{R}\backslash\{0\},\, \rho \in \mathbb{R}. \qquad (2.2.56)$$

Define the representation $\hat{\pi}_{\rho,\varepsilon}$ ($0 < |\rho| < 1$, $\varepsilon = 0$ or 1) of $SL(2,\mathbb{R})$, acting on functions defined on the real line, by the equation

$$\left(\hat{\pi}_{\rho,\varepsilon}(\left(\begin{smallmatrix} a & b \\ c & d \end{smallmatrix}\right)) w \right)(\sigma) = |-b\sigma + d|_\varepsilon^{-1-\rho}\, w \left(\frac{a\sigma - c}{-b\sigma + d} \right): \qquad (2.2.57)$$

when $\varepsilon = 0$, this is a representation taken from the complementary series of $SL(2,\mathbb{R})$; when $\varepsilon = 1$, it is a signed version, non unitarizable, of the same. More details can be found in [38, Sect. 2], with the same notation.

Proposition 2.2.12. *Under the map $u \mapsto ((\mathcal{Q}u)_0, (\mathcal{Q}u)_1)$, the anaplectic representation transfers to the representation $(\hat{\pi}_{-\frac{1}{2},0}, \hat{\pi}_{\frac{1}{2},1})$.*

Proof. Though it is contained in the above given reference, let us at least give a short indication about one of the possible proofs of the proposition. When $g = \left(\begin{smallmatrix} 1 & 0 \\ c & 1 \end{smallmatrix}\right)$, Ana$(g)$ is the multiplication by $e^{i\pi c x^2}$, and it is trivial to verify that $(\mathcal{Q}\,\mathrm{Ana}(g)\,u)_j(\sigma) = (\mathcal{Q}u)_j(\sigma - c)$; when $g = \left(\begin{smallmatrix} a & 0 \\ 0 & a^{-1} \end{smallmatrix}\right)$ with $a > 0$, so that $(\mathrm{Ana}(g)\,u)(x) = a^{-\frac{1}{2}} u(a^{-1}x)$, one verifies just as easily that

$$(\mathcal{Q}\,\mathrm{Ana}(g)\,u)(\sigma) = \left(a^{\frac{1}{2}} (\mathcal{Q}u)_0(a^2 \sigma), a^{\frac{3}{2}} (\mathcal{Q}u)_1(a^2 \sigma) \right). \qquad (2.2.58)$$

The case when $g = \left(\begin{smallmatrix} 0 & 1 \\ -1 & 0 \end{smallmatrix}\right)$ is of course more complicated, but there are several ways of dealing with it. Considering, say, the first component of the $\mathcal{Q}$-realization, one may prove instead the more general formula

$$\left(\mathcal{Q}\,\mathrm{Ana}\left(\left(\begin{smallmatrix} \cos t & \sin t \\ -\sin t & \cos t \end{smallmatrix}\right)\right)u\right)_0(\sigma) = |\cos t - \sigma \sin t|^{-\frac{1}{2}}\,(\mathcal{Q}u)_0\left(\frac{\sigma \cos t + \sin t}{-\sigma \sin t + \cos t}\right):$$

(2.2.59)

the advantage is that it is equivalent to its infinitesimal version, which takes (2.2.29) into account

$$-\frac{1}{2}\left[\left(\mathcal{Q}(Q^2 + P^2)u\right)_0(\sigma)\right] = \frac{1}{2i\pi}\left[\frac{\sigma}{2}(\mathcal{Q}u)(\sigma) + (1+\sigma^2)(\mathcal{Q}u)'(\sigma)\right]:$$

(2.2.60)

now, the left-hand side of (2.2.60), to wit

$$\int_{-\infty}^{\infty} e^{-\pi\sigma x^2}\left[\frac{ix^2}{2}u(xe^{-\frac{i\pi}{4}}) + \frac{1}{8\pi^2}u''(xe^{-\frac{i\pi}{4}})\right]dx, \qquad \sigma \text{ large}, \qquad (2.2.61)$$

can be written, after an integration by parts, as

$$\frac{1}{2i\pi}\int_{-\infty}^{\infty} e^{-\pi\sigma x^2}u(xe^{-\frac{i\pi}{4}})\left[-\pi(1+\sigma^2)x^2 + \frac{\sigma}{2}\right]dx, \qquad (2.2.62)$$

which is the right-hand side of the desired formula (2.2.60).   □

Remark 2.2.1. As is well known, the representation $\hat{\pi}_{\frac{1}{2},0}$, taken from the comple mentary series of $SL(2,\mathbb{R})$, is unitary for the scalar product associated to the norm such that

$$\|w\|^2_{-\frac{1}{2},0} = \int_{-\infty}^{\infty} \bar{w}(\sigma)\,(|D|^{\frac{1}{2}}w)(\sigma)\,d\sigma, \qquad (2.2.63)$$

where $|D|^{\frac{1}{2}}$ stands for the operator of convolution by the Fourier transform of the function $s \mapsto |s|^{\frac{1}{2}}$. Then, if $u \in \mathfrak{A}_{\mathrm{even}}$, $(u\,|\,u)$, as defined in (2.2.11), coincides with $\|(\mathcal{Q}u)_0\|^2_{-\frac{1}{2},0}$. Something similar holds with the odd part of $\mathfrak{A}$ – but one is then only dealing with a *pseudoscalar* product – trading the integral on the right-hand side of (2.2.63) for the one obtained when replacing $|s|^{\frac{1}{2}}$ by $|s|^{-\frac{1}{2}}$ sign s.

In anaplectic analysis, however, one cannot do much with the Hilbert completion of the space $\mathfrak{A}_{\mathrm{even}}$: it is, indeed, essential to use only functions on the line extending as entire functions, so as to take advantage of the relation (2.2.2) between the components of the $\mathbb{C}^4$-realization of u.

Chapter 3
The One-Dimensional Alternative Pseudodifferential Analysis

In this chapter, we introduce and study alternative pseudodifferential analysis, i.e., pseudodifferential analysis in connection with anaplectic analysis on the line. One of its most characteristic features is that it splits into an *ascending* and a quite similar *descending* parts: we shall concentrate on the first one. Under any operator from the ascending calculus, an eigenstate of the (standard or not) harmonic oscillator L_z transforms into the sum of a series of eigenstates of L_z with higher energy level. Section 3.1 introduces a formal definition of the ascending calculus and proves its covariance properties, a true but not "manifest" one when Heisenberg representation is concerned: in a remark at the end of the same section, we shall explain the geometric ideas that led to this definition.

Section 3.2 is devoted to more technical facts, including a characterization of appropriate classes of operators by properties of their symbols: just like Beals' characterization [2] in usual pseudodifferential analysis, it specifies classes of operators by properties of their iterated brackets with the infinitesimal operators of Heisenberg's representation. This will make it possible to study the composition of operators of the ascending calculus in Sect. 3.4. The most important result, in this direction, is that the sharp composition # of two symbols – this is always, in pseudodifferential analysis, the operation on symbols that corresponds to the composition of operators – can be made explicit in terms related to the Rankin–Cohen brackets of holomorphic functions on the upper half-plane.

The construction of the ascending calculus is based on the use of the family (A_z) of lowering operators, the formal analogues, in anaplectic analysis, of the annihilation operators from the usual analysis: more precisely, in this calculus, operators are built as integral superpositions of powers of the A_z's with *negative* exponents: indeed, in anaplectic analysis, every operator A_z is a linear automorphism of $\mathfrak{A}$, a point which will be rechecked in a detailed way in Sect. 3.3. This is in striking contrast with the case in usual analysis, in which no analogue of such a calculus can exist: Sect. 4.2 will provide more details.

3.1 Ascending Pseudodifferential Analysis

We need to consider the Schwartz space $S(\mathbb{R}^2)$ of rapidly decreasing C^∞ functions on the plane, and its isotypic subspaces $S_m(\mathbb{R}^2)$, a function lying in that space if it transforms as indicated in (2.1.8) under the group of rotations. If $h = \sum_{m \in \mathbb{Z}} h_m$, $h_m \in S_m(\mathbb{R}^2)$, the terms h_m of this decomposition are to be called the *isotypic* components of h. Besides, we must introduce a certain subspace of $S(\mathbb{R}^2)$, only needed when dealing with the Heisenberg representation and the associated covariance rule.

Definition 3.1.1. We shall denote as $S^A(\mathbb{R}^2)$ the space of functions h in $S(\mathbb{R}^2)$ which have the following property: the function $h(x, \xi)$ is the restriction to $\mathbb{R}^2$ of a function holomorphic in $\mathbb{C}^2$, denoted by the same letter; moreover, given $(\alpha, \beta) \in \mathbb{C}^2$, the function $T_{\alpha, \beta} h$ defined as

$$(T_{\alpha, \beta} h)(x, \xi) = e^{-2\pi\beta(x - i\xi)} h(x - i\alpha, \xi - \alpha) \tag{3.1.1}$$

again lies in $S(\mathbb{R}^2)$ and remains in a bounded subset of this space as long as $|\alpha| + |\beta|$ is bounded. The superscript A is a reference to the lowering operator in (2.2.8), as opposed to A^*: the space $S^A(\mathbb{R}^2)$ is not invariant under the symmetry $(x, \xi) \mapsto (x, -\xi)$. Note that $T_{\alpha, \beta} T_{\alpha', \beta'} = T_{\alpha + \alpha', \beta + \beta'}$.

This definition is given so that, setting $w = x + i\xi$, one should be able to exponentiate the operator $\frac{\partial}{\partial w} = \frac{1}{2}\left(\frac{\partial}{\partial x} - i\frac{\partial}{\partial \xi}\right)$ and the operator of multiplication by $\bar{w}$: indeed, one has

$$\frac{d}{dt} T_{t\alpha, t\beta} = \left(-2\pi\beta\,\bar{w} - 2i\alpha\,\frac{\partial}{\partial w}\right) T_{t\alpha, t\beta}. \tag{3.1.2}$$

For instance, all usual (two-dimensional) Hermite functions lie in $S^A(\mathbb{R}^2)$.

This space is invariant under rotations in the plane: indeed, if $(\mathcal{R}_\theta h)(x, \xi) = h(x\cos\theta - \xi\sin\theta, x\sin\theta + \xi\cos\theta)$, one has

$$\left(T_{\alpha, \beta} \mathcal{R}_\theta h\right)(x, \xi) = \left(\mathcal{R}_\theta T_{\alpha e^{i\theta}, \beta e^{i\theta}} h\right)(x, \xi). \tag{3.1.3}$$

Since $\sup_{|\beta| = R} \left| e^{-2\pi\beta(x - i\xi)} \right| = e^{2\pi R\sqrt{x^2 + \xi^2}}$, the Fourier transform of a function in the space under study again extends as a holomorphic function on $\mathbb{C}^2$. We now verify that the space $S^A(\mathbb{R}^2)$ remains invariant under the two-dimensional metaplectic representation of $SL(2, \mathbb{R})$. Using (2.1.6), it is just a matter of checking the formulas

$$T_{\alpha, \beta}\left(e^{i\pi c(x^2 + \xi^2)} h\right) = e^{i\pi c(x^2 + \xi^2)} T_{\alpha, \beta - cy} h,$$

$$T_{\alpha, \beta}\left(\mathcal{F} h\right) = \mathcal{F}\left(T_{-\beta, \alpha} h\right),$$

$$T_{\alpha, \beta}\left(H_a h\right) = H_a\left(T_{a\alpha, a\beta} h\right): \tag{3.1.4}$$

in the last equation, we have set $(H_a h)(x, \xi) = a^{-1} h(a^{-1}x, a^{-1}\xi)$.

The preceding equations show that the group of operators on $\mathcal{S}^A(\mathbb{R}^2)$ obtained with the help of the representation $\mathrm{Met}^{(2)}$ of $SL(2,\mathbb{R})$ normalizes the group made up by the transformations $T_{\alpha,\beta}$. In view of the covariance properties of the symbolic calculus to be introduced presently, this corresponds to the fact that, as pointed out in Theorem 2.2.7, it is possible to combine the anaplectic representation with the Heisenberg representation as a representation of the appropriate semidirect product.

We also denote as $\mathcal{S}_m^A(\mathbb{R}^2)$ the intersection $\mathcal{S}^A(\mathbb{R}^2) \cap L_m^2(\mathbb{R}^2)$, i.e., the mth iso-typic subspace of $\mathcal{S}^A(\mathbb{R}^2)$. It may seem obvious, but in our opinion it deserves a proof, that if $h \in \mathcal{S}_m^A$, one can write $h(q,p) = (q-ip)^m k(q^2+p^2)$ with k extending as an entire function of q^2+p^2. If $m=1$, we set $\beta(p) = \frac{\partial h}{\partial q}(0,p)$: the equation $h(-q,-p) = -h(q,p)$ shows that β is an even function, so that $\beta(p) = k(p^2)$ for some entire function k. In view of the equation $q\frac{\partial h}{\partial p} - p\frac{\partial h}{\partial q} = -ih$, one has $h(0,p) = -ip\frac{\partial h}{\partial q}(0,p) = -ipk(p^2)$, and the behavior of h under rotations then shows the identity $h(q,p) = (q-ip)k(q^2+p^2)$. To prove our claim for general m, it suffices to show (cf. Proposition 3.1.6 for another motivation) that if $h \in \mathcal{S}_{m+1}^A(\mathbb{R}^2)$ with $m \geq 1$, one can find h_1 and $h_2 \in \mathcal{S}_m^A(\mathbb{R}^2)$ such that, with $w = q+ip$, $h = i\bar{w}h_1 + \frac{1}{\pi}\frac{\partial h_2}{\partial w}$. Since $(m+1)h = -w\frac{\partial h}{\partial w} + \bar{w}\frac{\partial h}{\partial \bar{w}}$, it works with $h_1 = -\frac{i}{m}\frac{\partial h}{\partial \bar{w}}$ and $h_2 = -\frac{\pi}{m}wh$.

Finally, we set $(\mathcal{S}(\mathbb{R}^2))^\uparrow = \oplus_{m\geq 1} \mathcal{S}_m(\mathbb{R}^2)$ and $(\mathcal{S}^A(\mathbb{R}^2))^\uparrow = \oplus_{m\geq 1} \mathcal{S}_m^A(\mathbb{R}^2)$. If h lies in the latter space, so does $T_{\alpha,\beta}h$ for every (α,β) because, if $h(x,\xi) = (x-i\xi)^m k(x^2+\xi^2)$ with k extending as an entire function, the function

$$(T_{\alpha,\beta}h)(x,\xi) = e^{-2\pi\beta(x-i\xi)}(x-i\xi)^m k(x^2+\xi^2 - 2i\alpha(x-i\xi)) \qquad (3.1.5)$$

is an entire function of the pair of variables $(x^2+\xi^2, x-i\xi)$.

The definition that follows has a purely heuristic role and will be modified presently. It has been obtained at the end of a lengthy process, and we shall not attempt to give an a priori justification of it.

Let Q and P, respectively, stand for the usual position and momentum operators on the line, i.e., the multiplication by the variable x and the operator $\frac{1}{2i\pi}\frac{d}{dx}$. Given $h \in (\mathcal{S}(\mathbb{R}^2))^\uparrow = \oplus_{m\geq 1}\mathcal{S}_m(\mathbb{R}^2)$, one defines the operator

$$\mathrm{Op}^{\mathrm{asc}}(h) = \frac{1}{4\pi}\int_{\mathbb{R}^2}\int_\Pi (x+i\xi)\,h(x,\xi)\,e^{-\frac{i\pi}{z}(x^2+\xi^2)}\,dx\,d\xi$$

$$\left[x+i\xi - \frac{z}{\mathrm{Im}\,z}(Q-\bar{z}P)\right]^{-2}d\mu(z). \qquad (3.1.6)$$

To make this integral meaningful, we first decompose h as a sum $h = \sum_{m\geq 1}h_m$, with $h_m \in \mathcal{S}_m(\mathbb{R}^2)$, and analyze separately the operators corresponding to the iso-typic components of h. We start from the identity

$$\sum_{m\geq 1}mw^m t^{m+1} = w(w-t^{-1})^{-2}, \qquad (3.1.7)$$

in which we set $w = x + i\xi$ and substitute for t the operator

$$\frac{\operatorname{Im} z}{z}(Q - \bar{z}P)^{-1} = \pi^{\frac{1}{2}}\frac{\operatorname{Im} z}{z}A_z^{-1}. \tag{3.1.8}$$

Of course, this operator is meaningless in usual analysis, since, in this context, $(\operatorname{Im} z)^{-\frac{1}{2}}A_z$ is the conjugate, under some element of the one-dimensional meta-plectic transformation, of the annihilation operator $A = \pi^{\frac{1}{2}}(Q + iP)$, certainly not an invertible operator. However, in anaplectic analysis, this operator makes perfectly good sense: this could be guessed from Theorem 2.2.3, though some estimates are required [38, p. 163]. Proposition 3.3.1 will contain a totally different proof of this fact, a fundamental one for our purposes. Since (cf. Remark 2.2.1) we do not want to substitute, say for $\mathfrak{A}_{\mathrm{even}}$, its Hilbert completion, the series on the left-hand side of (3.1.7) only converges in a weak sense, for instance, as will be shown in Lemma 3.2.1, the following: given any pair z, ζ of points of Π, as a series of operators on the linear space generated by eigenfunctions of the (anaplectic) harmonic oscillator L_ζ, valued in the algebraic dual of that space. However, for the time being, we shall satisfy ourselves with using the new form of the defining equation (3.1.6), rewritten as

$$\operatorname{Op}^{\mathrm{asc}}(h) = \frac{1}{4\pi}\sum_{m \geq 1}m\,\pi^{\frac{m+1}{2}}\int_\Pi \left(\frac{\operatorname{Im} z}{z}\right)^{m+1}A_z^{-m-1}\,d\mu(z)$$
$$\int_{\mathbb{R}^2}w^m h(x,\xi)\,e^{-\frac{i\pi}{z}|w|^2}\,dx\,d\xi, \tag{3.1.9}$$

as a definition: in the last integral, we may substitute h_m for h without change. Taking Proposition 2.1.1 into consideration, we finally state our definition as follows.

Definition 3.1.2. Given $h = \sum_{m \geq 1}h_m \in (\mathcal{S}(\mathbb{R}^2))^\uparrow$, we set

$$\operatorname{Op}^{\mathrm{asc}}(h) = \sum_{m \geq 1}\operatorname{Op}_m^{\mathrm{asc}}(h_m), \tag{3.1.10}$$

where $\operatorname{Op}_m^{\mathrm{asc}}$ is defined on the space $\mathcal{S}_m(\mathbb{R}^2)$ by the equation

$$\operatorname{Op}_m^{\mathrm{asc}}(h) = \frac{m}{4\pi}\pi^{\frac{m+1}{2}}\int_\Pi (\Theta_m h)(z)\,A_z^{-m-1}\,(\operatorname{Im} z)^{m+1}\,d\mu(z). \tag{3.1.11}$$

Postponing the questions of convergence, we first show that the calculus so defined is covariant under the following pair of representations of $SL(2, \mathbb{R})$: the anaplectic representation Ana acting on functions of one variable and the representation $\operatorname{Met}^{(2)}$ acting on symbols. Since the latter one preserves the decomposition $(\mathcal{S}(\mathbb{R}^2))^\uparrow = \oplus_{m \geq 1}\mathcal{S}_m(\mathbb{R}^2)$, this is really a statement concerning each calculus $\operatorname{Op}_m^{\mathrm{asc}}$.

Remark 3.1.1. Looking back at Proposition 2.1.1, one sees why it is necessary to consider only summands h_m with $m \geq 1$ (or $m \leq -1$), so that the map $\operatorname{Op}^{\mathrm{asc}}$ will have a chance to be one to one. Much more precise indications will be given in Remark 3.2.1(iii).

With the help of (2.2.22), one obtains

$$\text{Ana}\left(\begin{pmatrix} a & b \\ c & d \end{pmatrix}\right) \text{Op}_m^{\text{asc}}(h) \, \text{Ana}\left(\begin{pmatrix} d & -b \\ -c & a \end{pmatrix}\right)$$

$$= \frac{m}{4\pi} \pi^{\frac{m+1}{2}} \int_\Pi (\Theta_m h)(z) \, (c\bar{z} + d)^{-m-1} A_{\frac{az+b}{cz+d}}^{-m-1} (\text{Im } z)^{m+1} \, d\mu(z)$$

$$= \frac{m}{4\pi} \pi^{\frac{m+1}{2}} \int_\Pi (\Theta_m h)(\frac{dz - b}{-cz + a})(-cz + a)^{-m-1} A_z^{-m-1} (\text{Im } z)^{m+1} \, d\mu(z), \quad (3.1.12)$$

the same as $\text{Op}_m^{\text{asc}}\left(\text{Met}^{(2)}(g) h\right)$ according to Proposition 2.1.1.

The proof of covariance under the (anaplectic) Heisenberg representation is more difficult and requires a number of lemmas.

Lemma 3.1.3. *For $m = 1, 2, \dots$ and $h \in S_m(\mathbb{R}^2)$, one has*

$$\Theta_m\left(|w|^2 h\right) = \frac{1}{i\pi} [z^2 \frac{d}{dz} + (m+1) z] \, \Theta_m h. \quad (3.1.13)$$

Proof. Since

$$|w|^2 e^{-\frac{i\pi}{z}|w|^2} = \frac{z^2}{i\pi} \frac{d}{dz} (e^{-\frac{i\pi}{z}|w|^2}), \quad (3.1.14)$$

onc has

$$\Theta_m\left(|w|^2 h\right)(z) = \frac{1}{i\pi} z^{-m+1} \frac{d}{dz} \int_{\mathbb{R}^2} w^m e^{-\frac{i\pi}{z}|w|^2} h(x, \xi) \, dx d\xi$$

$$= \frac{1}{i\pi} [z^2 \frac{d}{dz} + (m+1) z] \left(z^{-m-1} \int_{\mathbb{R}^2} w^m e^{-\frac{i\pi}{z}|w|^2} h(x, \xi) \, dx d\xi \right).$$

$$(3.1.15)$$

$\square$

Lemma 3.1.4. *Under the assumptions of the lemma that precedes, one has*

$$\Theta_{m+1}(i\pi \bar{w} h) = (z\frac{d}{dz} + m + 1) \Theta_m h,$$

$$\Theta_{m+1}(\frac{\partial h}{\partial w}) = \frac{d}{dz} \Theta_m h. \quad (3.1.16)$$

Proof. One has

$$(\Theta_{m+1}(i\pi \bar{w} h))(z) = z^{-m-2} \int_{\mathbb{R}^2} w^{m+1} e^{-\frac{i\pi}{z}|w|^2} i\pi \bar{w} h(x, \xi) \, dx d\xi$$

$$= \frac{i\pi}{z} \Theta_m\left(|w|^2 h\right)(z) \quad (3.1.17)$$

and one obtains the first relation from an application of Lemma 3.1.3. Next,

$$
\begin{aligned}
\left(\Theta_{m+1}\left(\frac{\partial h}{\partial w}\right)\right)(z) &= z^{-m-2} \int_{\mathbb{R}^2} w^{m+1} e^{-\frac{i\pi}{z}|w|^2} \frac{\partial h}{\partial w}(x,\xi)\, dx d\xi \\
&= -z^{-m-2} \int_{\mathbb{R}^2} \frac{\partial}{\partial w} \left(w^{m+1} e^{-\frac{i\pi}{z}|w|^2} \right) h(x,\xi)\, dx d\xi \\
&= z^{-m-2} \int_{\mathbb{R}^2} e^{-\frac{i\pi}{z}|w|^2} w^m \left[\frac{i\pi |w|^2}{z} - (m+1) \right] h(x,\xi)\, dx d\xi \\
&= z^{-2} [z^2 \frac{d}{dz} + (m+1)z] (\Theta_m h)(z) - (m+1) z^{-1} (\Theta_m h)(z)
\end{aligned}
$$
(3.1.18)

according to Lemma 3.1.3.   □

Lemma 3.1.5. *For $m \geq 1$,*

$$
[Q, A_z^{-m-1}] = -\frac{m+1}{2i\pi^{\frac{1}{2}}} \bar{z} A_z^{-m-2},
$$

$$
[P, A_z^{-m-1}] = -\frac{m+1}{2i\pi^{\frac{1}{2}}} A_z^{-m-2}.
$$
(3.1.19)

Proof. Since $A_z = \pi^{\frac{1}{2}} (Q - \bar{z} P)$, it is immediate by induction that, for $m \geq 1$,

$$
[Q, A_z^m] = \frac{m}{2i\pi^{\frac{1}{2}}} \bar{z} A_z^{m-1}, \qquad [P, A_z^m] = \frac{m}{2i\pi^{\frac{1}{2}}} A_z^{m-1} :
$$
(3.1.20)

the lemma follows.   □

Next, we note that, setting $D_0 = \frac{d}{dz}$ and $D_1 = z\frac{d}{dz} - m$, one has

$$
\bar{z}^j (\operatorname{Im} z)^{m-1} = \frac{2i}{m} D_j (\operatorname{Im} z)^m.
$$
(3.1.21)

In view of some integrations by parts, we locally prefer to use on Π the Lebesgue measure $dV(z) = d\operatorname{Re} z\, d\operatorname{Im} z = (\operatorname{Im} z)^2 d\mu(z)$: with respect to it, the transpose of the operators D_0 and D_1 are

$$
D_0^\top = -\frac{d}{dz}, \qquad D_1^\top = -z\frac{d}{dz} - (m+1).
$$
(3.1.22)

We are now in a position to prove the following.

Proposition 3.1.6. *Given $h \in S_m(\mathbb{R}^2)$, one has*

$$
[Q, \operatorname{Op}_m^{\mathrm{asc}}(h)] = \operatorname{Op}_{m+1}^{\mathrm{asc}} (i\bar{w} h),
$$

$$
[P, \operatorname{Op}_m^{\mathrm{asc}}(h)] = \operatorname{Op}_{m+1}^{\mathrm{asc}} (\frac{1}{\pi} \frac{\partial h}{\partial w}).
$$
(3.1.23)

Proof. From Lemma 3.1.5, one has (do not confuse $d\mu$ and dV in what follows)

$$[Q, \mathrm{Op}_m^{\mathrm{asc}}(h)] = \frac{m}{4\pi} \pi^{\frac{m+1}{2}} \int_\Pi (\Theta_m h)(z) [Q, A_z^{-m-1}] (\mathrm{Im}\, z)^{m+1} d\mu(z)$$

$$= -\frac{m(m+1)}{8i\pi^{\frac{3}{2}}} \pi^{\frac{m+1}{2}} \int_\Pi (\Theta_m h)(z) A_z^{-m-2} \bar{z} (\mathrm{Im}\, z)^{m-1} dV(z). \quad (3.1.24)$$

We now use the case ($j = 1$) of (3.1.21) and immediately transpose the operator D_1, transforming what precedes into

$$-\frac{(m+1)}{4\pi} \pi^{\frac{m}{2}} \int_\Pi (D_1^\top \Theta_m h)(z) A_z^{-m-2} (\mathrm{Im}\, z)^{m+2} d\mu(z) : \quad (3.1.25)$$

now, according to Lemma 3.1.4,

$$-(D_1^\top \Theta_m h)(z) = (z \frac{d}{dz} + m + 1)(\Theta_m h)(z)$$

$$= (\Theta_{m+1}(i\pi \bar{w} h))(z), \quad (3.1.26)$$

so that

$$[Q, \mathrm{Op}_m^{\mathrm{asc}}(h)] = \frac{m+1}{4\pi} \pi^{\frac{m}{2}} \int_\Pi [\Theta_{m+1}(i\pi \bar{w} h)](z) A_z^{-m-2} (\mathrm{Im}\, z)^{m+2} d\mu(z)$$

$$= \frac{1}{\pi} \mathrm{Op}_{m+1}^{\mathrm{asc}}(i\pi \bar{w} h). \quad (3.1.27)$$

When P is substituted for Q, the only difference is that one must use D_0 in place of D_1, which leads to the second equation (3.1.23). □

The pair of equations (3.1.23) can be written as

$$\left.\frac{d}{dt}\right|_{t=0} e^{2i\pi t (\eta Q - yP)} \mathrm{Op}^{\mathrm{asc}}(h) e^{-2i\pi t (\eta Q - yP)} = 2i\pi [\eta Q - yP, \mathrm{Op}^{\mathrm{asc}}(h)]$$

$$= \mathrm{Op}^{\mathrm{asc}}\left((-2\pi \eta \bar{w} - 2iy \frac{\partial}{\partial w}) h\right). \quad (3.1.28)$$

Let us sum up the results obtained.

Theorem 3.1.7. *The ascending pseudodifferential analysis satisfies the two covariance properties expressed by the equations*

$$\mathrm{Ana}\left(\left(\begin{smallmatrix} a & b \\ c & d \end{smallmatrix}\right)\right) \mathrm{Op}^{\mathrm{asc}}(h) \, \mathrm{Ana}\left(\left(\begin{smallmatrix} d & -b \\ -c & a \end{smallmatrix}\right)\right) = \mathrm{Op}^{\mathrm{asc}}\left(\mathrm{Met}^{(2)}\left(\left(\begin{smallmatrix} a & b \\ c & d \end{smallmatrix}\right)\right) h\right), \quad (3.1.29)$$

in which it is assumed that $h \in (\mathcal{S}(\mathbb{R}^2))^\uparrow$, and, in the case when $h \in (\mathcal{S}^A(\mathbb{R}^2))^\uparrow$,

$$e^{2i\pi(\eta Q - yP)} \operatorname{Op}^{\mathrm{asc}}(h) e^{-2i\pi(\eta Q - yP)} = \operatorname{Op}^{\mathrm{asc}}(T_{y,\eta} h), \tag{3.1.30}$$

where the representation T of $\mathbb{R}^2$ in $S^A(\mathbb{R}^2)$ has been defined in the beginning of the present section.

Proof. The first equation has already been mentioned; the second one is a consequence of the exponentiated version

$$e^{2i\pi(\eta Q - yP)} \operatorname{Op}^{\mathrm{asc}}(h) e^{-2i\pi(\eta Q - yP)} = \operatorname{Op}^{\mathrm{asc}}\left(\exp\left(-2\pi\eta\,\bar{w} - 2iy\frac{\partial}{\partial w}\right)h\right) \tag{3.1.31}$$

of (3.1.28) and of (3.1.2).   □

Theorem 3.1.8 will not be of much use in what follows. However, it completes our understanding that, in the usual and alternative pseudodifferential analyses, the Euler operator $\mathcal{E} = (2i\pi)^{-1}(x\frac{\partial}{\partial x} + \xi\frac{\partial}{\partial \xi} + 1)$ and the rotation operator $\mathcal{R} = \xi\frac{\partial}{\partial x} - x\frac{\partial}{\partial \xi}$ play dual roles. In [36, p. 120], we introduced the *mixed adjoint* operation $\operatorname{mad}\Lambda$, defined, in the case when $\Lambda = P \wedge Q$, as

$$(\operatorname{mad}\Lambda)T = PTQ - QTP: \tag{3.1.32}$$

we then observed the following formula, valid in the Weyl calculus:

$$(\operatorname{mad}\Lambda)\operatorname{Op}(\mathfrak{S}) = \operatorname{Op}(\mathcal{E}\,\mathfrak{S}). \tag{3.1.33}$$

This operation turned out to play an essential role in automorphic pseudodifferential analysis since [36, p. 144] it made it possible to kill the pole at $\lambda = 0$ of the spectral density of the sharp composition of two Eisenstein distributions. Let us just see what is the analogue, in the ascending pseudodifferential analysis, of (3.1.33).

Theorem 3.1.8. *Given* $h \in (S(\mathbb{R}^2))^\dagger$, *one has*

$$(\operatorname{mad}(P \wedge Q))\operatorname{Op}^{\mathrm{asc}}(h) = \frac{1}{\pi}\operatorname{Op}^{\mathrm{asc}}(\mathcal{R}h). \tag{3.1.34}$$

Proof. For every $z \in \Pi$, the operators Q and P are linear combinations of A_z and A_z^*, respectively: the simple formulas can be found in (3.2.35): applying these formulas, one finds

$$2i\pi(\operatorname{mad}(P \wedge Q))A_z^{-m-1} = \frac{2}{\operatorname{Im} z}[A_z^{-m}, A_z^*]. \tag{3.1.35}$$

Now, as observed in Proposition 2.2.8, $[A_z, A_z^*] = \operatorname{Im} z$: it follows (by induction for $m \geq 1$, then for m of any sign) that $[A_z^m, A_z^*] = m\operatorname{Im} z A_z^{m-1}$, hence

$$2i\pi(\operatorname{mad}(P \wedge Q))A_z^{-m-1} = -2mA_z^{-m-1}. \tag{3.1.36}$$

The theorem follows in view of the fact that, if $h \in S_m(\mathbb{R}^2)$, one has $\mathcal{R}h = imh$. □

Before we come to a brief description of the origin of the definition, we wish to give a rough argument to the effect that nothing can be changed in Definition 3.1.2, save for changing $\frac{1}{4\pi}$ (but not m) to another factor, if we insist on preserving the covariance properties of the calculus. Indeed, let Op^1 and Op^2 be two symbolic calculi with $\mathbb{R}^2$ as a phase space, both satisfying the covariance properties in Theorem 3.1.7, and let us concentrate on the operators $\text{Met}^{(2)}(g)$ and $T_{y,\eta}$ which occur on the right-hand sides. We also assume that, on some space of symbols including $(\mathcal{S}^A(\mathbb{R}^2))^\uparrow$, the linear operators Op^1 and Op^2 are one to one, so that a formula like

$$\text{Op}^2(h) = \text{Op}^1(\Lambda h) \tag{3.1.37}$$

holds for some appropriate operator Λ on $(\mathcal{S}^A(\mathbb{R}^2))^\uparrow$. Then, Λ has to commute with all operators $\text{Met}^{(2)}(g)$ and $T_{y,\eta}$. These are operators associated with representations, so that we may consider the corresponding infinitesimal operators: ultimately, looking at the list (2.1.6) and at (3.1.2), we obtain that the operator Λ must commute with the operator of multiplication by the function $x^2 + \xi^2 = |w|^2$, with the Fourier transformation, with the multiplication by $\bar{w}$, finally with the differential operators $w\frac{\partial}{\partial w} + \bar{w}\frac{\partial}{\partial \bar{w}}$ and $\frac{\partial}{\partial w}$. Then, at least when considered on functions "divisible" by $\bar{w}$ in the space $(\mathcal{S}^A(\mathbb{R}^2))^\uparrow$, Λ must commute with the operators of multiplication by w or $\bar{w}$: it has to be an operator of multiplication, actually by a constant since it commutes with the Fourier transformation. Note that a fully similar argument proves the uniqueness of the Weyl calculus, if its two basic covariance properties are to be satisfied.

Still, we shall most of the time pay more attention to the first covariance relation (3.1.29) than to (3.1.30). It is the only one likely to play a role in possible applications to modular form theory.

Remark 3.1.2. At one referee's urging, we give some brief indications about the lengthy process that led to Definition 3.1.2. The present remark is not logically needed for the sequel, but the geometrical ideas on which it is based may interest some readers, as similar ones may resurface in related situations.

The first aim was to build a symbolic calculus satisfying the covariance relation (3.1.29) as well as *some* covariance relation involving the Heisenberg representation: in view of the conjugation relation (2.2.19), there was not much choice left (nothing more than a unique normalizing constant) so far as the right-hand sides of (3.1.23) are concerned.

Next, the quantization theory of homogeneous spaces of $G = SL(2,\mathbb{R})$ [32, 34] – this means making such a homogeneous space the phase space of a symbolic calculus of operators acting on the space of an *irreducible* representation π of G – leads to the following conclusion, in conformity with Kirillov's method of orbits. In the case when π is taken from the projective discrete series (resp., from the principal, or complementary, series, let us say in general from some part of the full nonunitary principal series [16]), a good choice of phase space can be the orbit $\mathfrak{H}$ (resp. $\mathfrak{H}_i$) of the coadjoint action of G in the dual $\mathfrak{g}^*$ of the Lie algebra $\mathfrak{g}$ of G, where $\mathfrak{H}$ is one sheet of a two-sheeted hyperboloid and $\mathfrak{H}_i$ is the one-sheeted hyperboloid defined by the equation $x_0^2 - x_1^2 - x_2^2 = -1$ in appropriate coordinates. In association with the coadjoint action of G on such an orbit, consider the quasiregular representation of G on $L^2(\mathfrak{H})$ or $L^2(\mathfrak{H}_i)$. It is extremely classical that the first one decomposes as an integral sum of representations from the principal series only while, from results of Strichartz [30], the second one decomposes into a continuous and a discrete part, the second one being the sum of a series of representations from the discrete series.

Since our project, based on the search for the validity of (3.1.29), claims for using spaces of symbols decomposing into (subspaces of) Hilbert spaces associated to the discrete series, it became clear that we had better start with a representation π taken from the full nonunitary principal series of G. In view of letting the Heisenberg representation play its part, it was clear too that, rather than using for π an irreducible representation, we had to use some related composite object. At this point, the anaplectic representation shows itself as being the good candidate, being, as indicated by Proposition 2.2.12, the direct sum of two irreducible representations of G, the first one the representation $\hat{\pi}_{-\frac{1}{2},0}$ from the complementary series, the second one, $\hat{\pi}_{\frac{1}{2},1}$, a signed (non unitarizable) version of the same. Finding a symbolic calculus of operators, covariant under the anaplectic representation, quite appropriately started from the consideration of a calculus of operators based on the use of $\mathfrak{H}_i$ as a temporary phase space.

The corresponding quantization program, including a study of the sharp composition of symbols, was implemented in [34]; the quantizing map, in that case, had independently been obtained in [22]. In the present investigations, special consideration is attached to the discrete part only of the decomposition of $L^2(\mathfrak{H}_i)$. It consists of two series, the first of which is equivalent to the sum of all terms $\mathcal{D}_{2n+2}$ (cf. (2.1.11)) with $n = 0, 1, \dots$ and the second is the complex conjugate of the first. The nth space from the first sum is the closure, in $L^2(\mathfrak{H}_i)$, of the space of functions $x \mapsto g_z^{n+1}(x) = (\langle \xi, x \rangle)^{-n-1}$, where $z \in \Pi$ and $\xi = (\frac{1}{2}(1 + \bar{z}^2), \frac{1}{2}(1 - \bar{z}^2), -\bar{z})$, a point of $\mathfrak{g}_{\mathbb{C}}$. Specializing in the representation $\hat{\pi}_{-\frac{1}{2},0}$, one sees from the results of [34] that the operator on the line with symbol g_z^{n+1} is some explicit multiple of the operator D_z^{-n-1}, where D_z is the differential operator such that

$$-4i\pi D_z = (\sigma - \bar{z})\left[(\sigma - \bar{z})\frac{d}{d\sigma} + \frac{1}{2}\right]$$

$$= \left[(\sigma - \bar{z})\frac{d}{d\sigma} - \frac{1}{2}\right](\sigma - \bar{z}) \qquad (3.1.38)$$

in terms of the coordinate σ on the line (cf.(2.2.57)): yes, the operator D_z is invertible, when regarded as an endomorphism of the space of C^∞ vectors of the representation $\hat{\pi}_{-\frac{1}{2},0}$.

On the other hand, the symbols g_z^{n+1}, $z \in \Pi$, define the integral kernel of the intertwining operator W_{n+1} from the space $\mathcal{H}_{2n+2}$ of holomorphic functions in Π (recall Proposition 2.1.1) to the corresponding subspace from the decomposition of the discrete part of $L^2(\mathfrak{H}_i)$ since [34, p. 126] (forgetting again the normalization constants)

$$(W_{n+1}f)(x) = \int_\Pi f(z)\, g_z^{n+1}(x)\, (\mathrm{Im}\, z)^{2n+2}\, d\mu(z), \qquad x \in \mathfrak{H}_i. \tag{3.1.39}$$

Combining the map W_{n+1} with the symbolic calculus with $\mathfrak{H}_i$ as a phase space, one obtains a linear map, satisfying of course the desired covariance property, from the space $\mathcal{H}_{2n+2}$ to a space of operators, to wit

$$f \mapsto \int_\Pi f(z)\, D_z^{-n-1}\, (\mathrm{Im}\, z)^{2n+2}\, d\mu(z): \tag{3.1.40}$$

taking Proposition 2.1.1 into account, one may also regard this map as defined on the space $\mathcal{S}_{2n+1}(\mathbb{R}^2)$.

We now use the identification of the Hilbert space of the representation $\hat{\pi}_{-\frac{1}{2},0}$ with a completion of the even part of the space $\mathfrak{A}$, as provided (Proposition 2.2.12) by the map $u \mapsto (Qu)_0$ in (2.2.54). Actually, we did not use quite the same definition of the full nonunitary series in [34] and in (2.2.57): the operators obtained are (when $(\rho,\varepsilon) = (-\frac{1}{2},0)$) the conjugate of one another under the involution θ such that $(\theta w)(\sigma) = |\sigma|^{-\frac{1}{2}} w(\sigma^{-1})$. After some computation, one finds that, under the transfer $u \mapsto \theta\, (Qu)_0$, the operator D_z transforms to the restriction to $\mathfrak{A}_{\mathrm{even}}$ of the operator iA_z^2.

This leads to (3.1.11), with the understanding that the necessity (and uniqueness) of such a definition is only established for odd m, up to some coefficient *depending on m*, and only so far as the action on $\mathfrak{A}_{\mathrm{even}}$ is concerned. There is no natural *absolute* normalization of the calculus since the identity operator does not fit within it. However, having chosen the normalizing constant corresponding to $m = 1$, the other ones are fully determined since equation

$$[P,[P,A_z^{-2n-2}]] = -\frac{(2n+2)(2n+3)}{4\pi}\, A_z^{-2n-4}, \tag{3.1.41}$$

a consequence of (3.1.19), together with the second equation (3.1.23), the desirability of which has been established in the beginning of the present remark, connects the pieces of the calculus corresponding to values $2n+1$ and $2n+3$ of m. We now have a calculus, acting on $\mathfrak{A}_{\mathrm{even}}$ only, with $\oplus_{m=1,3,\dots} \mathcal{S}_m(\mathbb{R}^2)$ as a space of symbols, which satisfies the covariance equations (3.1.29) as well as the "second-order" equations (3.1.23), i.e., those involving, just like (3.1.41), two commutators with operators from the pair (Q, P) taken in succession. Actually, things work just as well, with the same normalization constants, so far as the action on $\mathfrak{A}_{\mathrm{odd}}$ is concerned (the commutation relations do not feel the difference). However, it should be clear that, up to now, we only have a symbolic calculus of operators which preserve each of the two subspaces of $\mathfrak{A}$ with a definite parity. Also, it only uses the sum of subspaces $\mathcal{S}_m(\mathbb{R}^2)$ of $\mathcal{S}(\mathbb{R}^2)$ with odd m as a space of symbols. To complete our calculus, we still have to use the sum of subspaces $\mathcal{S}_m(\mathbb{R}^2)$ with even m and make it a space of symbols for operators changing the parity of functions. The "first-order" equations (3.1.23) leave no room for choice.

3.2 Classes of Operators

We are now ready to start with the more technical matters.

Lemma 3.2.1. *Given* $m = 0, 1, \ldots,\ j, k \in \mathbb{Z}$ *and* $z,\ \zeta \in \Pi$, *the pseudoscalar product* $(A_z^{-m-1}\,\phi_\zeta^k \mid \phi_\zeta^j)$ *is given by the equation*

$$(A_z^{-m-1}\,\phi_\zeta^k \mid \phi_\zeta^j) = \overline{C_m^{j,k}}\,(\mathrm{Im}\,\zeta)^{\frac{m+1}{2}}\,(z-\zeta)^{\frac{-m-1+j-k}{2}}\,(z-\bar{\zeta})^{\frac{-m-1-j+k}{2}} \qquad (3.2.1)$$

for some constant $C_m^{j,k}$.

One has $C_m^{j,k} = 0$ *unless* $m+1-j+k$ *is even and* $m+1 \le j-k$. *As a special case,*

$$C_m^{k+m+1,k} = \begin{cases} (-2i)^{m+1}\,2^{-2k}\,\dfrac{(2k)!}{k!} & \text{if } k \ge 0, \\[2mm] (-2i)^{m+1} & \text{if } -m \le k \le -1, \\[2mm] (-2i)^{m+1}\,(-1)^{k+m+1}\,2^{2k+2m+2}\,\dfrac{|2k+2m+2|!}{|k+m+1|!} & \text{if } k \le -m-1. \end{cases} \qquad (3.2.2)$$

Proof. Using (2.2.23), one has

$$(A_z^{-m-1}\,\phi_\zeta^k \mid \phi_\zeta^j) = (\mathrm{Ana}(g_\zeta^{-1})\,A_z^{-m-1}\,\mathrm{Ana}(g_\zeta)\,\phi^k \mid \phi^j) \qquad (3.2.3)$$

with $g_\zeta^{-1} = \begin{pmatrix} \eta^{-\frac{1}{2}} & -\eta^{-\frac{1}{2}}\xi \\ 0 & \eta^{\frac{1}{2}} \end{pmatrix}$ if $\zeta = \xi + i\eta$; from (2.2.22),

$$\mathrm{Ana}(g_\zeta^{-1})\,A_z^{-m-1}\,\mathrm{Ana}(g_\zeta) = \left(\eta^{\frac{1}{2}}\,A_{\frac{z-\xi}{\eta}} \right)^{-m-1} : \qquad (3.2.4)$$

(3.2.3) and (3.2.4) reduce the proof of the lemma to the case when $\zeta = i$, which we assume from now on.

Set $H_m^{j,k}(z) = (A_z^{-m-1}\,\phi^k \mid \phi^j)$, a holomorphic function of z (recall from (2.2.11) that the pseudoscalar product is antilinear with respect to its argument on the left). From Proposition 2.2.8 again, one finds that, for every $g = \begin{pmatrix} a & b \\ c & d \end{pmatrix} \in SL(2,\mathbb{R})$, one has

$$H_m^{j,k}\left(\frac{az+b}{cz+d}\right) = (cz+d)^{m+1}\,(A_z^{-m-1}\,\mathrm{Ana}(g^{-1})\,\phi^k \mid \mathrm{Ana}(g^{-1})\,\phi^j) : \qquad (3.2.5)$$

in particular, taking for g the matrix $k_\theta = \begin{pmatrix} \cos\frac{\theta}{2} & -\sin\frac{\theta}{2} \\ \sin\frac{\theta}{2} & \cos\frac{\theta}{2} \end{pmatrix}$ and using (2.2.30), to the effect that

$$\mathrm{Ana}(k_\theta^{-1})\,\phi^j = \exp\left(-\frac{i\theta}{2}L\right)\phi^j = e^{-\frac{ij\theta}{2}}\,\phi^j, \qquad (3.2.6)$$

we obtain that

$$H_m^{j,k}\left(\frac{z\cos\frac{\theta}{2}-\sin\frac{\theta}{2}}{z\sin\frac{\theta}{2}+\cos\frac{\theta}{2}}\right) = e^{\frac{i(k-j)\theta}{2}}(z\sin\frac{\theta}{2}+\cos\frac{\theta}{2})^{m+1}H_m^{j,k}(z). \qquad (3.2.7)$$

Considering this equation for $\theta = 2\pi$, one sees that this is impossible, if $H_m^{j,k}$ is not zero, unless $m+1-j+k$ is an even number. Set, for $z \neq i$,

$$G_m^{j,k}(z) = (z-i)^{\frac{m+1-j+k}{2}}(z+i)^{\frac{m+1+j-k}{2}}H_m^{j,k}(z): \qquad (3.2.8)$$

one verifies that

$$G_m^{j,k}\left(\frac{z\cos\frac{\theta}{2}-\sin\frac{\theta}{2}}{z\sin\frac{\theta}{2}+\cos\frac{\theta}{2}}\right) = G_m^{j,k}(z), \qquad (3.2.9)$$

from which it follows that $G_m^{j,k}$ is a constant and that $j-k-m-1 \geq 0$ if $H_m^{j,k} \neq 0$.

What remains to be done is finding, in the case when $j-k=m+1$, the constant in front of the right-hand side of (3.2.1). Fix $w=i$. Using (2.2.34), one obtains

$$A^{-1}\phi^k = \phi^{k+1} \text{ if } k \leq -1, \qquad A^{-1}\phi^k = (k+\frac{1}{2})^{-1}\phi^{k+1} \text{ if } k \geq 0, \qquad (3.2.10)$$

so that, iterating the last pair of equations,

$$A^{-m-1}\phi^k = \begin{cases} 2^{2m+2}\frac{(2k)!}{k!}\frac{(k+m+1)!}{(2k+2m+2)!}\phi^{k+m+1} & \text{if } k \geq 0, \\ 2^{2m+2k+2}\frac{(k+m+1)!}{(2k+2m+2)!}\phi^{k+m+1} & \text{if } -m \leq k \leq -1, \quad (3.2.11) \\ \phi^{k+m+1} & \text{if } k \leq -m-1. \end{cases}$$

Then, the scalar product $(\phi^{k+m+1}\,|\,\phi^{k+m+1})$ is to be found in (2.2.12), and one obtains the constant in (3.2.2) from a look at the special case of (3.2.1) for which $z = w = i$. $\qquad \Box$

So far as we know, there is no simple formula for the coefficients $C_m^{j,k}$ in general: however, there are several recurrence relations, and Lemma 3.2.2 makes a recursive computation feasible.

Lemma 3.2.2. *With the notation of Lemma 3.2.1, set*

$$C_m^{j,k} = \frac{(-2i)^{m+1}}{m!}F_m^{j,k}, \qquad (3.2.12)$$

a number which can be nonzero only if $j-k-m-1$ is even and nonnegative: in the case when $j-k-m-1=0$, this number has been made explicit in (3.2.2). For $j-k-m-1 = 2,4,\ldots$, one has

$$F_m^{j,k} = \frac{1}{j-k-m-1}[2\gamma_j F_{m+1}^{j-1,k} - F_{m+2}^{j,k}], \qquad (3.2.13)$$

which makes it possible to compute $F_m^{j,k}$ by induction relative to the nonnegative integer $\frac{j-k-m-1}{2}$.

Proof. By induction,

$$[P, (Q - \bar{z}P)^{-k}] = -\frac{k}{2i\pi} (Q - \bar{z}P)^{-k-1}. \tag{3.2.14}$$

Next,

$$\frac{\partial}{\partial \bar{z}} (Q - \bar{z}P)^{-1} = (Q - \bar{z}P)^{-1} P (Q - \bar{z}P)^{-1} \tag{3.2.15}$$

and

$$\frac{\partial}{\partial \bar{z}} (Q - \bar{z}P)^{-m} = mP (Q - \bar{z}P)^{-m-1} + \frac{m(m+1)}{4i\pi} (Q - \bar{z}P)^{-m-2}, \tag{3.2.16}$$

so that

$$\frac{\partial}{\partial \bar{z}} A_z^{-m-1} = (m+1)\,\pi^{\frac{1}{2}}\,P A_z^{-m-2} + \frac{(m+1)(m+2)}{4i} A_z^{-m-3} \tag{3.2.17}$$

and

$$\frac{\partial}{\partial z} H_m^{j,k} = (m+1)\,\pi^{\frac{1}{2}}\,(P A_z^{-m-2} \phi^k | \phi^j) - \frac{(m+1)(m+2)}{4i} (A_z^{-m-3} \phi^k | \phi^j). \tag{3.2.18}$$

Since

$$\pi^{\frac{1}{2}} P = (\bar{z} + i)^{-1} (A - A_z) \tag{3.2.19}$$

and $A^* \phi^j = \gamma_j^* \phi^{j+1}$, one has

$$\pi^{\frac{1}{2}} (P A_z^{-m-2} \phi^k | \phi^j) = (z - i)^{-1} \left[\gamma_j^* (A_z^{-m-2} \phi^k | \phi^{j+1}) - (A_z^{-m-1} \phi^k | \phi^j) \right],$$
$$\tag{3.2.20}$$

so that

$$\frac{\partial}{\partial z} H_m^{j,k} = (m+1)(z-i)^{-1} [\gamma_j^* H_{m+1}^{j+1,k} - H_m^{j,k}] - \frac{(m+1)(m+2)}{4i} H_{m+2}^{j,k} \tag{3.2.21}$$

and, using the expression (3.2.1) of the functions involved,

$$i(m+1+j-k) C_m^{j,k} + (m+1)\gamma_j^* C_{m+1}^{j+1,k} = -\frac{(m+1)(m+2)}{4i} C_{m+2}^{j,k}. \tag{3.2.22}$$

On the other hand, using

$$A_z^* = \frac{i}{2} (z - i) A - \frac{i}{2} (z + i) A^* \tag{3.2.23}$$

and (2.2.34), one obtains

$$H_m^{j,k} = (A_z^{-m-2}\phi^k | A_z^* \phi^j)$$

$$= \frac{i}{2}(z-i)(A_z^{-m-2}\phi^k | A\phi^j) - \frac{i}{2}(z+i)(A_z^{-m-2}\phi^k | A^*\phi^j)$$

$$= \frac{i}{2}\gamma_j (z-i) H_{m+1}^{j-1,k} - \frac{i}{2}\gamma_j^* (z+i) H_{m+1}^{j+1,k} : \tag{3.2.24}$$

using (3.2.1) again, this implies

$$C_m^{j,k} - \frac{i}{2}\gamma_j^* C_{m+1}^{j+1,k} = -\frac{i}{2}\gamma_j C_{m+1}^{j-1,k}. \tag{3.2.25}$$

Solving (3.2.22) and (3.2.25) for the unknown constants on the left-hand sides, one obtains, if $j - k - m - 1 \geq 2$,

$$C_m^{j,k} = \frac{m+1}{-m-1+j-k}\left[\frac{m+2}{4}C_{m+2}^{j,k} + i\gamma_j C_{m+1}^{j-1,k}\right]. \tag{3.2.26}$$

Since $j - k - m - 1$ decreases by two units when moving from the left-hand side to the right-hand side, this equation makes it possible to obtain, at least inductively, the constants $C_m^{j,k}$ (recall that, unless in the zero case, the number $j - k - m - 1$ is even and nonnegative), starting from the value of $C_m^{k+m+1,k}$ indicated in (3.2.2).

In terms of the coefficients $F_m^{j,k}$, (3.2.26) becomes

$$F_m^{j,k} = \frac{1}{j-k-m-1}[2\gamma_j F_{m+1}^{j-1,k} - F_{m+2}^{j,k}]. \tag{3.2.27}$$

The following application of Lemma 3.2.2 will be helpful later.

Lemma 3.2.3. *Let m be an integer ≥ 1 and let $\mu = [\frac{m-1}{2}]$. Let $k_0, \ldots, k_\mu$ be $\mu + 1$ pairwise distinct nonnegative integers. Then, the vectors*

$$\xi_v = \begin{pmatrix} F_m^{k_v+m+1,k_v} \\ F_{m-2}^{k_v+m+1,k_v} \\ \cdots \\ F_{m-2\mu}^{k_v+m+1,k_v} \end{pmatrix}, \qquad 0 \leq v \leq \mu, \tag{3.2.28}$$

make up a linear basis of $\mathbb{C}^{\mu+1}$.

Proof. For any given pair (m,r) with $0 \leq r \leq \mu = [\frac{m-1}{2}]$, the coefficient $F_{m-2r}^{k+m+1,k}$ is the product of $2^{-2k}\frac{(2k)!}{k!}$ by a polynomial $P_r(k)$ of degree r exactly. Indeed, if $r = 0$, this is true by the definition (3.2.12) of the coefficients $F_m^{j,k}$, to be completed by (3.2.2). If $r \geq 1$, one can rewrite (3.2.27), with the help of (2.2.35), as

$$F_{m-2r}^{k+m+1,k} = \frac{1}{r}\left[2(k+m+\frac{1}{2})F_{m-1-2(r-1)}^{k+(m-1)+1,k} - F_{m-2(r-1)}^{k+m+1,k}\right], \tag{3.2.29}$$

which proves our claim by induction on r. Replacing ξ_v by the proportional vector $\eta_v = 2^{2k_v}\frac{k_v!}{(2k_v)!}\xi_v$, we must show that the determinant

$$
\begin{vmatrix}
P_0 & \cdots & P_0 \\
P_1(x_0) & \cdots & P_1(x_\mu) \\
\cdots & \cdots & \cdots \\
P_\mu(x_0) & \cdots & P_\mu(x_\mu)
\end{vmatrix}
\tag{3.2.30}
$$

is not zero if the numbers $x_0, \ldots, x_\mu$ are pairwise distinct: by the usual Vandermonde trick, this determinant is the product of a constant by the polynomial $\prod_{j<k}(x_j - x_k)$, and the constant is not zero since the coefficient of $x_1 x_2^2 \ldots x_\mu^\mu$ in the expansion of the determinant is the product of the top-order coefficients of the polynomials $P_0, \ldots, P_\mu$. $\square$

It is convenient, at least for a temporary purpose, to introduce the space $(L^2_{\text{weak}}(\mathbb{R}^2))^\uparrow$ consisting of formal series $h = \sum_{m \geq 1} h_m$, with $h_m \in L^2_m(\mathbb{R}^2)$ for every m. In view of applying Proposition 3.1.6, we shall also consider the space $(S_{\text{weak}}(\mathbb{R}^2))^\uparrow$ consisting of formal series $h = \sum_{m \geq 1} h_m$, with $h_m \in S_m(\mathbb{R}^2)$; this is equivalent to saying that h_m lies in $L^2_m(\mathbb{R}^2)$ and satisfies the following property: with $w = x + i\xi$, the function $\bar{w}^\alpha \left(\frac{\partial}{\partial w}\right)^\beta h_m$ lies in $L^2_{m+\alpha+\beta}(\mathbb{R}^2)$ for every pair of nonnegative integers α, β. Finally, we shall say that a formal series $\sum_{m \geq 1} h_m$ lies in $(S'_{\text{weak}}(\mathbb{R}^2))^\uparrow$ if, for every $m \geq 1$, h_m is a linear combination of distributions $\bar{w}^\alpha \left(\frac{\partial}{\partial w}\right)^\beta f_{m-\alpha-\beta}$ with $\alpha + \beta < m - 1$ and $f_{m-\alpha-\beta} \in I^2_{m-\alpha-\beta}(\mathbb{R}^2)$: because of the condition $\alpha + \beta \leq m - 1$, this space is strictly smaller than the subspace of $S'(\mathbb{R}^2)$ consisting in distributions the isotypic components of which transform according to (2.1.8) under the action of the rotation group.

If $h \in (L^2_{\text{weak}}(\mathbb{R}^2))^\uparrow$, the operator $\text{Op}^{\text{asc}}(h)$ has at least a very weak meaning, toward the characterization of which we introduce, for every $\zeta \in \Pi$, the linear space E_ζ generated by all functions ϕ^j_ζ with $\zeta \in \Pi$ and $j \in \mathbb{Z}$, and its algebraic dual E'_ζ.

Lemma 3.2.4. *Given $h = \sum_{m \geq 1} h_m \in (L^2_{\text{weak}}(\mathbb{R}^2))^\uparrow$ and any point $\zeta \in \Pi$, the operator $\sum_{m \geq 1} \text{Op}^{\text{asc}}_m(h_m)$ makes sense as a linear operator from E_ζ to E'_ζ.*

Proof. As a preparation, the result of which will also be useful elsewhere, we compute, for $\zeta \in \Pi$, a certain Θ_m-transform: assuming $m \geq 1$ and using the definition in Proposition 2.1.1,

$$
\left[\Theta_m \left((x - i\xi)^m \exp\left(\frac{i\pi}{\zeta}(x^2 + \xi^2)\right)\right)\right](z)
\tag{3.2.31}
$$

$$
= z^{-m-1} \int_{\mathbb{R}^2} (x^2 + \xi^2)^m \exp\left(-i\pi\left(\frac{1}{z} - \frac{1}{\zeta}\right)(x^2 + \xi^2)\right) dx d\xi
$$

$$
= 2\pi z^{-m-1} \int_0^\infty t^{2m+1} \exp\left(-i\pi\left(\frac{1}{z} - \frac{1}{\zeta}\right)t^2\right) dt
$$

$$
= m! \, \pi^{-m} i^{m+1} \left[\frac{z - \bar{\zeta}}{\zeta}\right]^{-m-1}.
\tag{3.2.32}
$$

In particular, the function $z \mapsto (z - \bar{\zeta})^{-m-1}$ is square integrable, on Π, with respect to the measure $(\mathrm{Im}\, z)^{m+1}\, d\mu(z)$. So is a function such as

$$(\bar{z} - \bar{\zeta})^{\frac{j-k-m-1}{2}} (\bar{z} - \zeta)^{\frac{k-j-m-1}{2}} = (\bar{z} - \zeta)^{-m-1} \times \left[\frac{\bar{z} - \bar{\zeta}}{\bar{z} - \zeta}\right]^{\frac{j-k-m-1}{2}} \tag{3.2.33}$$

in the case when $j - k - m - 1$ is a nonnegative even integer, since in this case the second factor on the right-hand side remains bounded for $z, \zeta \in \Pi$.

Using (3.1.11) and taking advantage of Lemma 3.2.1, one obtains, given $j, k \in \mathbb{Z}$, that

$$\sum_{m \geq 1} (\phi_\zeta^j \,|\, \mathrm{Op}_m^{\mathrm{asc}}(h_m)\, \phi_\zeta^k)$$

$$= \sum_{m=1}^{j-k-1} \frac{m}{4\pi} \pi^{\frac{m+1}{2}} \int_\Pi (\Theta_m h_m)(z)\, (\phi_\zeta^j \,|\, A_z^{-m-1}\, \phi_\zeta^k)\, (\mathrm{Im}\, z)^{m+1}\, d\mu(z)$$

$$= \sum_{m=1}^{j-k-1} \frac{m}{4\pi} \pi^{\frac{m+1}{2}} C_m^{j,k} (\mathrm{Im}\, \zeta)^{\frac{m+1}{2}} \times$$

$$\int_\Pi (\Theta_m h_m)(z)\, (\bar{z} - \bar{\zeta})^{\frac{-m-1+j-k}{2}} (\bar{z} - \zeta)^{\frac{-m-1-j+k}{2}} (\mathrm{Im}\, z)^{m+1}\, d\mu(z). \tag{3.2.34}$$

The main point is that this is only a finite sum; next, the function $(\Theta_m h_m)(z)$ is square integrable with respect to the measure $(\mathrm{Im}\, z)^{m+1}\, d\mu(z)$ according to Proposition 2.1.1, and so is the second factor under the last integral sign according to (3.2.33), so that the Cauchy–Schwarz inequality gives a meaning to the right-hand side of (3.2.34). □

It is easy to deal with linear operators: $E_\zeta \to E_\zeta'$ because, within this class, one can always multiply any operator, on the left or on the right, by any linear combination of Q and P. This is so in view of the formulas

$$Q = \pi^{-\frac{1}{2}} \frac{\zeta A_\zeta - \bar{\zeta} A_\zeta^*}{2i \,\mathrm{Im}\, \zeta}, \qquad P = \pi^{-\frac{1}{2}} \frac{A_\zeta - A_\zeta^*}{2i \,\mathrm{Im}\, \zeta}, \tag{3.2.35}$$

to be completed by (2.2.34).

We may now extend the weak definition of $\mathrm{Op}^{\mathrm{asc}}(h)$ to the case when $h \in (\mathcal{S}'_{\mathrm{weak}}(\mathbb{R}^2))^\dagger$, in such a way that Proposition 3.1.6 should continue to hold. One obvious point is that, since in this case h_m is a tempered distribution for every $m \geq 1$, the definition of $\Theta_m h_m$ extends, and that Lemma 3.1.4 is still valid. Let us set (the first item is just a notational convenience)

$$\delta_0^{m+1} = \frac{1}{\pi} \frac{\partial}{\partial z}, \qquad \delta_1^{m+1} = \frac{1}{\pi} (z \frac{\partial}{\partial z} + m + 1). \tag{3.2.36}$$

What we want to do is to *define* $B = \mathrm{Op}^{\mathrm{asc}}\left((i\bar{w})^{\alpha} \left(\frac{1}{\pi} \frac{\partial}{\partial w} \right)^{\beta} f_{m-\alpha-\beta} \right)$ as

$$B = [Q, [Q, \ldots [P, [P, \ldots, \mathrm{Op}^{\mathrm{asc}}_{m-\alpha-\beta}(f_{m-\alpha-\beta})]\ldots]\ldots], \qquad (3.2.37)$$

where the number of Q's is α and the number of P's is β. Indeed, according to Proposition 3.1.6, this equation is correct in the case when $f_{m-\alpha-\beta} \in S_m(\mathbb{R}^2)$. Using Lemma 3.1.4 and (3.2.34), setting $\chi_{m-\alpha-\beta+1} = \Theta_{m-\alpha-\beta} h_{m-\alpha-\beta}$ and considering an arbitrary sequence $(\varepsilon_1, \ldots, \varepsilon_{\alpha+\beta})$ of 0's and 1's such that $\sum \varepsilon_r = \alpha$, we may write, in this case,

$$(\phi^j_\zeta | B \phi^k_\zeta) = \frac{1}{4\pi} \pi^{\frac{m+1}{2}} (\mathrm{Im}\, \zeta)^{\frac{m+1}{2}} \int_\Pi \left(\delta^m_{\varepsilon_1} \ldots \delta^{m-\alpha-\beta+1}_{\varepsilon_{\alpha+\beta}} \chi_{m-\alpha-\beta+1} \right)(z)$$
$$(\bar{z} - \bar{\zeta})^{\frac{-m-1+j-k}{2}} (\bar{z} - \zeta)^{\frac{-m-1-j+k}{2}} (\mathrm{Im}\, z)^{m-1}\, dV(z), \quad (3.2.38)$$

with the understanding that the right-hand side should be set to zero unless $j - k - m - 1$ is an even nonnegative integer: recall from the proof of Lemma 3.1.5 that dV is the Lebesgue measure on $\Pi \subset \mathbb{R}^2$.

However, to use this as a definition, we must show that, under the sole assumption that $f_{m-\alpha-\beta} \in L^2_{m-\alpha-\beta}(\mathbb{R}^2)$, the right-hand side of (3.2.38) can still be given a meaning, depending only on the function $\delta^m_{\varepsilon_1} \ldots \delta^{m-\alpha-\beta+1}_{\varepsilon_{\alpha|\beta}} \chi_{m\ \alpha\ \beta|1}$ rather than on $\chi_{m-\alpha-\beta+1}$. The integral under consideration is not convergent in general: let $\kappa \in C^\infty(\mathbb{R})$ have a compact support and satisfy $\kappa(0) = 1$, $\kappa'(0) = \kappa''(0) = \cdots = 0$. Recalling that the hyperbolic distance $d(i,z)$, in Π, is given by the equation $\cosh d(i,z) = \frac{1+|z|^2}{2\,\mathrm{Im}\, z}$, we shall prove that the (convergent) integral obtained, from that on the right-hand side of (3.2.38), by inserting the factor $\kappa(t \cosh d(i,z))$, with $t > 0$, under the integral sign, has a limit as $t \to 0$, to wit the convergent integral

$$\int_\Pi \chi_{m-\alpha-\beta+1}(z)(\delta^{m-\alpha-\beta+1}_{\varepsilon_{\alpha+\beta}})^\top \ldots (\delta^m_{\varepsilon_1})^\top (\mathrm{Im}\, z)^{m-1}$$
$$\times (\bar{z} - \bar{\zeta})^{\frac{-m-1+j-k}{2}} (\bar{z} - \zeta)^{\frac{-m-1-j+k}{2}}\, dV(z), \quad (3.2.39)$$

where $(\delta^\gamma_\varepsilon)^\top$ indicates the transpose of the operator $\delta^\gamma_\varepsilon$ with respect to dV: recall that the action of these on powers of $\mathrm{Im}\, z$ has already been indicated in (3.1.21) and (3.1.22). One shows by induction on r that, for every $r = 0, 1, \ldots$, the function

$$(\delta^{m-r+1}_{\varepsilon_r})^\top \ldots (\delta^m_{\varepsilon_1})^\top \left[(\mathrm{Im}\, z)^{m-1} \kappa(t \cosh d(i,z)) \right] \qquad (3.2.40)$$

can be written as a linear combination of terms of the kind

$$R(z, \bar{z}) (\mathrm{Im}\, z)^{m-s-1} = \left(\frac{1 + \bar{z}^2}{\mathrm{Im}\, z} \right)^p P(\bar{z})\, t^q\, \kappa^{(q)}(t \cosh d(i,z)) (\mathrm{Im}\, z)^{m-s-1},$$
$$(3.2.41)$$

where P is a polynomial of degree $\leq s$ and $q \leq r$, $s \leq r$, $p \leq q$: besides, the only term with $q = 0$ reduces, as is immediate, to

$$\kappa\left(t\cosh d(i,z)\right)\times\left(\delta_{\mathcal{E}_r}^{m-r+1}\right)^{\top}\dots\left(\delta_{\mathcal{E}_1}^{m}\right)^{\top}(\operatorname{Im}z)^{m-1}. \tag{3.2.42}$$

To prove this, we have to plug in an extra $\left(\delta_{\mathcal{E}_{r+1}}^{m-r}\right)^{\top}$ on the left and check what happens. This extra operator is either $-\frac{1}{\pi}\frac{\partial}{\partial z}$ or $-\frac{1}{\pi}\left(z\frac{\partial}{\partial z}-m+r+1\right)$. In the second case, we split the image, under this operator, of $R(z,\bar{z})\,(\operatorname{Im}z)^{m-s-1}$, as

$$R(z,\bar{z})\left(z\frac{\partial}{\partial z}-m+s+1\right)(\operatorname{Im}z)^{m-s-1}+\left(z\frac{\partial}{\partial z}+r-s\right)(R(z,\bar{z}))\cdot(\operatorname{Im}z)^{m-s-1}. \tag{3.2.43}$$

We then use (3.1.21) to the effect that

$$\left(z\frac{\partial}{\partial z}-m+s+1\right)(\operatorname{Im}z)^{m-s-1}=\frac{m-s-1}{2i}\,\bar{z}\,(\operatorname{Im}z)^{m-s-2}. \tag{3.2.44}$$

The rest is a question of care only, writing

$$\frac{\partial}{\partial z}\left(\kappa^{(q)}(t\cosh d(i,z))\right)=\frac{it}{4}\left(\frac{1+\bar{z}^2}{\operatorname{Im}z}\right)\kappa^{(q+1)}(t\cosh d(i,z))\times(\operatorname{Im}z)^{-1},$$

$$\frac{\partial}{\partial z}\left(\frac{1+\bar{z}^2}{\operatorname{Im}z}\right)^{p}=\frac{ip}{2}\left(\frac{1+\bar{z}^2}{\operatorname{Im}z}\right)^{p}\times(\operatorname{Im}z)^{-1} \tag{3.2.45}$$

and, when needed, $z=\bar{z}+2i\operatorname{Im}z$.

Next, we have to prove that every integral

$$\int_{\Pi}\chi_{m-\alpha-\beta+1}(z)\left(\frac{1+\bar{z}^2}{\operatorname{Im}z}\right)^{p}P(\bar{z})\,t^{q}\,\kappa^{(q)}(t\cosh d(i,z))\,(\operatorname{Im}z)^{m-s-1}$$

$$(\bar{z}-\bar{\zeta})^{\frac{-m-1+j-k}{2}}(\bar{z}-\zeta)^{\frac{-m-1-j+k}{2}}\,dV(z), \tag{3.2.46}$$

with p,q,r satisfying the constraints above in terms of $r=\alpha+\beta$ and $\deg P\leq s$, has a well-defined limit as $t\to0$, to wit the integral (3.2.39) in the case when $q=0$ (in which case it is assumed, as justified in (3.2.42), that $s=r$, $q=p=0$, and $P=1$), and zero if $q>0$.

Using the dominated convergence theorem, all we have to do, since the function $\left(\frac{1+\bar{z}^2}{\operatorname{Im}z}\right)^{p}t^{q}\,\kappa^{(q)}(t\cosh d(i,z))$ is uniformly bounded for small t as $p\leq q$, is showing the convergence of the integral

$$\int_{\Pi}|\chi_{m-\alpha-\beta+1}(z)|\,(1+|z|)^{s}\,|\bar{z}-\bar{\zeta}|^{\frac{-m-1+j-k}{2}}\,|\bar{z}-\zeta|^{\frac{-m-1-j+k}{2}}\,(\operatorname{Im}z)^{m-s-1}\,dV(z), \tag{3.2.47}$$

under the assumption that $-m-1+j-k$ is a nonnegative even number. At the end of the proof of Lemma 3.2.4, we checked the convergence of an integral such as

$$\int_{\Pi}|\chi_{m-\alpha-\beta+1}(z)|\,|\bar{z}-\bar{\zeta}|^{\frac{-m-1+\ell+\alpha+\beta}{2}}\,|\bar{z}-\zeta|^{\frac{-m+\alpha+\beta-1-\ell}{2}}\,(\operatorname{Im}z)^{m-\alpha-\beta-1}\,dV(z), \tag{3.2.48}$$

under the assumption that $\ell - (m - \alpha - \beta + 1)$ is a nonnegative even integer. In (3.2.47), we thus set $j - k = \ell + \alpha + \beta$, transforming this integral into an integral identical to (3.2.48), save for the extra factor

$$(1 + |z|)^s |\bar{z} - \zeta|^{-\alpha - \beta} (\operatorname{Im} z)^{\alpha + \beta - s} = \left(\frac{\operatorname{Im} z}{1 + |z|} \right)^{\alpha + \beta - s} \left(\frac{1 + |z|}{|\bar{z} - \zeta|} \right)^{\alpha + \beta}, \quad (3.2.49)$$

a bounded one since $s \leq r = \alpha + \beta$.

We have proved the following.

Proposition 3.2.5. *It is possible to associate to every $h \in (S'_{\text{weak}}(\mathbb{R}^2))^\uparrow$, in a linear way, an operator $\operatorname{Op}^{\text{asc}}(h)$ regarded as a collection of linear operators: $E_\zeta \to E'_\zeta$, so that the following conditions hold:*

(i) *in the case when $h \in (L^2_{\text{weak}}(\mathbb{R}^2))^\uparrow$, the definition agrees with that given in Lemma 3.2.4;*

(ii) *for any $h \in (S'_{\text{weak}}(\mathbb{R}^2))^\uparrow$, the equations (3.1.23) giving the symbols of the commutator of an operator from the ascending symbolic calculus with Q or P continue to hold.*

It will be handy to let the notion of weak integral over Π enter the picture: as suggested by the proof that precedes, it is by definition the limit as $t \to 0$, assumed to exist (in which case we say that the integral is weakly convergent), of the integral obtained when inserting the extra factor $\kappa(t \cosh d(i, z))$.

We shall now prove that the map $\operatorname{Op}^{\text{asc}}$ is one to one, by showing how to recover any given isotypic component h_m of a symbol $h \in (S'_{\text{weak}}(\mathbb{R}^2))^\uparrow$ from the knowledge, for appropriate finite sets of values of (j, k), of the scalar products $(\phi^j_\zeta | \operatorname{Op}^{\text{asc}}(h) \phi^k_\zeta)$ as functions of ζ.

Theorem 3.2.6. *Let $h \in (S'_{\text{weak}}(\mathbb{R}^2))^\uparrow$ and let $B = \operatorname{Op}^{\text{asc}}(h)$, as defined in a weak sense according to Proposition 3.2.5. Given $r \geq 1$, let*

$$\mathfrak{S}(r) = \{m \colon 1 \leq m \leq r; \ m \equiv r \bmod 2\}. \quad (3.2.50)$$

Given any finite sequence $X = (X_1, \ldots, X_\ell)$ of operators each of which coincides either with Q or with P, and any pair $j, k \in \mathbb{Z}$, all scalar products $(\phi^j_\zeta | [X_1, [X_2, \ldots [X_\ell, B] \ldots]] \phi^k_\zeta)$ are zero unless $j - k \geq \ell + 2$. If such is the case, one can uniquely decompose the family (depending on ζ) of scalar products so defined as a sum

$$(\phi^j_\zeta | [X_1, [X_2, \ldots, [X_\ell, B] \ldots]] \phi^k_\zeta) = \sum_{s-1 \in \mathfrak{S}(j-k-1)} (\operatorname{Im} \zeta)^{\frac{s}{2}} T^{j,k}_{X,s}(\zeta), \quad (3.2.51)$$

where the functions $T^{j,k}_{X,s}$ are holomorphic. When $\ell = 0$, we abbreviate $T^{j,k}_{X,s}$ as $T^{j,k}_s$. The map $\operatorname{Op}^{\text{asc}}$ is one to one.

In the case when $h \in (S_{\text{weak}}(\mathbb{R}^2))^{\uparrow}$, the function $T_{\mathbf{x},s}^{j,k}$ lies in the space $\mathcal{H}_s$ of holomorphic functions in Π introduced in Proposition 2.1.1 for every s.

Proof. We first consider the case when $h \in (S_{\text{weak}}(\mathbb{R}^2))^{\uparrow}$, dealing with the case $\ell = 0$ (no commutator is involved) of (3.2.51) to start with. Set $\chi_{m+1} = \Theta_m h_m$ (cf. Proposition 2.1.1) for simplicity of notation.

Starting from (3.2.34), let us recall from Lemma 3.2.1 that only the terms such that $m \in \mathfrak{S}(j-k-1)$ (cf. (3.2.50)) can contribute to this scalar product. Using the binomial expansion relative to the power of $\bar{z} - \bar{\zeta} = \bar{z} - \zeta + 2i\,\mathrm{Im}\,\zeta$, with an integral nonnegative exponent, which occurs in the integral, we obtain, changing locally m to n,

$$(\phi_\zeta^j \,|\, \mathrm{Op}^{\mathrm{asc}}(h)\, \phi_\zeta^k) = \sum_{n \in \mathfrak{S}(j-k-1)} \frac{n}{4\pi}\, \pi^{\frac{n+1}{2}}\, C_n^{j,k} \sum_{r=0}^{\frac{j-k-1-n}{2}} \binom{\frac{j-k-1-n}{2}}{r}$$

$$(\mathrm{Im}\,\zeta)^{\frac{n+1}{2}+r}\,(2i)^r \int_\Pi \chi_{n+1}(z)\,(\bar{z}-\zeta)^{-n-1-r}\,(\mathrm{Im}\,z)^{n+1}\,d\mu(z), \quad (3.2.52)$$

where the coefficient $C_n^{j,k}$ is the one defined in (3.2.1). One finds the expansion (3.2.51) of this function of ζ as a polynomial in $(\mathrm{Im}\,\zeta)^{\frac{1}{2}}$ with holomorphic coefficients if one sets $s = n + 1 + 2r$, a number in the finite set characterized by the condition that $s - 1 \in \mathfrak{S}(j-k-1)$: then, for any given s, the domain of possible n's is the set $\mathfrak{S}(s-1)$. The holomorphic coefficients $T_s^{j,k}$ are to be defined by the equation

$$T_s^{j,k}(\zeta) = \sum_{n \in \mathfrak{S}(s-1)} \frac{n}{4\pi}\, \pi^{\frac{n+1}{2}}\, C_n^{j,k} \binom{\frac{j-k-1-n}{2}}{\frac{s-n-1}{2}}$$

$$(2i)^{\frac{s-n-1}{2}} \int_\Pi \chi_{n+1}(z)\,(\bar{z}-\zeta)^{\frac{-n-1-s}{2}}\,(\mathrm{Im}\,z)^{n+1}\,d\mu(z). \quad (3.2.53)$$

In view of Bergman's reproducing kernel equation which, in the space $\mathcal{H}_{n+1}$, expresses itself as

$$\chi(\zeta) = \frac{n}{4\pi} \int_\Pi \left[\frac{i}{2}\,(\bar{z}-\zeta)\right]^{-n-1} \chi(z)\,(\mathrm{Im}\,z)^{n+1}\,d\mu(z), \quad (3.2.54)$$

the integral above can be written as

$$4\pi\left(\frac{i}{2}\right)^{n+1} \frac{(n-1)!}{\left(\frac{s+n-1}{2}\right)!}\, \chi_{n+1}^{\left(\frac{s-n-1}{2}\right)}(\zeta). \quad (3.2.55)$$

As seen from (3.2.53) to (3.2.55), saying that the function $T_s^{j,k}$ lies in the space $\mathcal{H}_s$ is the same as saying that the functions $\chi_{n+1}^{\left(\frac{s-n-1}{2}\right)}$ do. Since $\chi_{n+1} \in \mathcal{H}_{n+1}$ by definition, all that needs being proved is that the operator $\frac{\partial}{\partial z}$ maps any space $\mathcal{H}_{n+1}$

to $\mathcal{H}_{n+3}$. Now [34, p. 129] a form of Paley–Wiener theorem shows that the Laplace transform $\mathcal{L}_{n+1}$, normalized as

$$(\mathcal{L}_{n+1} u)(z) = \frac{(4\pi)^{\frac{n}{2}}}{((n-1)!)^{\frac{1}{2}}} \int_0^\infty u(\tau) e^{2i\pi z\tau} d\tau, \tag{3.2.56}$$

maps, in an isometric way, the space of functions u on $(0, \infty)$ such that $\|u\|_n^2 := \int_0^\infty |u(\tau)|^2 \tau^{-n} d\tau < \infty$ onto $\mathcal{H}_{n+1}$. It follows that, if $\chi \in \mathcal{H}_{n+1}$, the function $\chi' = \frac{\partial \chi}{\partial z}$ lies in $\mathcal{H}_{n+3}$ and one has $\|\chi'\|_{\mathcal{H}_{n+3}} = \frac{(n(n+1))^{\frac{1}{2}}}{2} \|\chi\|_{\mathcal{H}_{n+1}}$.

According to Proposition 3.2.5, substituting for an operator $B = \mathrm{Op}^{\mathrm{asc}}(h)$ a commutator $[X_1, [X_2, \ldots [X_\ell, B] \ldots]]$ with $X_r = Q$ or P for every r has the effect of substituting for the (formal) sum $\sum_{m \geq 1} h_m$ a sum of a similar kind, in which the subscript m only runs through values $\geq \ell + 1$. This implies that, on the right-hand side of (3.2.52), one can only obtain nonzero terms now when $j - k \geq \ell + 2$.

It will be handy to change notation slightly in the result of (3.2.53)–(3.2.55), setting $j = k + m + 1 + 2p$, $s = m + 1$ (so that $n = m - 2r$): with $\mu = [\frac{m-1}{2}]$, the result is

$$T_{m+1}^{k+m+1+2p,k}(\zeta)$$
$$= \sum_{r=0}^{\mu} \frac{(m-2r)!}{(m-r)!} \binom{r+p}{r} \pi^{\frac{m-2r+1}{2}} C_{m-2r}^{k+m+1+2p,k} (2i)^r \left(\frac{i}{2}\right)^{m-2r+1} \chi_{m+1-2r}^{(r)}(\zeta) \tag{3.2.57}$$

or, in terms of the constants $F_m^{j,k}$ introduced in Lemma 3.2.2,

$$T_{m+1}^{k+m+1+2p,k}(\zeta) = \sum_{r=0}^{\mu} \frac{(2i)^r}{(m-r)!} \binom{r+p}{r} \pi^{\frac{m-2r+1}{2}} F_{m-2r}^{k+m+1+2p,k} \chi_{m+1-2r}^{(r)}(\zeta). \tag{3.2.58}$$

Then,

$$(\phi_\zeta^j \,|\, \mathrm{Op}^{\mathrm{asc}}(h)\, \phi_\zeta^k) = \sum_{m \in \mathfrak{S}(j-k-1)} (\mathrm{Im}\ \zeta)^{\frac{m+1}{2}} T_{m+1}^{j,k}(\zeta). \tag{3.2.59}$$

This already shows that the map $\mathrm{Op}^{\mathrm{asc}}$, when regarded as defined on $(\mathcal{S}_{\mathrm{weak}}(\mathbb{R}^2))^\uparrow$, is one to one, since (3.2.58) makes it possible to recover χ_{m+1} (hence h_m) from the knowledge of any of the functions $T_{m+1}^{k+m+1,k}$ together with that of the functions χ_{m+1-2r} (or h_{m-2r}) for $1 \leq r \leq \mu$: of course, the functions $T_{m+1}^{k+m+1,k}$ are known if the operator $B = \mathrm{Op}^{\mathrm{asc}}(h)$ is.

However, we must now describe how all this can be extended to the case when it is only assumed that $h \in (\mathcal{S}'_{\mathrm{weak}}(\mathbb{R}^2))^\uparrow$. The modifications to be made are the following: first, extend (3.2.52), substituting for the integrals on the right-hand side the weak integrals, as explained just after Proposition 3.2.5, with the same integrands; next, extend Bergman's reproducing kernel equation (3.2.54). So far as the first question is concerned, the only point to check is that expanding

$(\bar{z} - \bar{\zeta})^{\frac{-m-1+j-k}{2}} = [(\bar{z} - \zeta) + 2i\,\mathrm{Im}\,\zeta]^{\frac{-m-1+j-k}{2}}$ by the binomial formula preserves the weak integrability of the terms obtained from the decomposition of the (now, weak) integral (3.2.34). Since the factor concerned here does not depend on z, just on $\bar{z}$, it is only necessary to remark (so as to extend the argument based on (3.2.33)) that, just like $\frac{\bar{z}-\bar{\zeta}}{\bar{z}-\zeta}$, the ratio $\frac{\mathrm{Im}\,\zeta}{\bar{z}-\zeta}$ is a bounded function of z, for a given ζ. The right-hand side of Bergman's reproducing kernel equation (3.2.54) has to be reinterpreted as a weak integral, in which $\chi_{n+1} \in \mathcal{H}_{n+1}$ is replaced by some function $\delta_{\varepsilon_1}^n \ldots \delta_{\varepsilon_r}^{n-r+1} \chi_{n-r+1}$ with $0 \leq r \leq n$ and $\chi_{n-r+1} \in \mathcal{H}_{n-r+1}$: we use the notation (3.2.36). Following the integration by parts and limiting process carried in (3.2.39)–(3.2.46), we end up with the interpretation of the weak integral substitute for the one on the right-hand side of (3.2.54) as a (genuine) integral

$$\frac{n}{4\pi} \int_{\Pi} \left[\frac{i}{2}(\bar{z} - \zeta)\right]^{-n-1} \chi_{n-r+1}(z)\,(\delta_{\varepsilon_r}^{n-r+1})^{\top} \ldots (\delta_{\varepsilon_1}^{n})^{\top} (\mathrm{Im}\,z)^{n-1}\,dV(z): \quad (3.2.60)$$

we must show that it coincides with

$$4\pi \left(\frac{i}{2}\right)^{n+1} (n-1)!\,(\delta_{\varepsilon_1}^n \ldots \delta_{\varepsilon_r}^{n-r+1} \chi_{n-r+1})(\zeta). \quad (3.2.61)$$

This is the case since it is true when $\delta_{\varepsilon_1}^n \ldots \delta_{\varepsilon_r}^{n-r+1} \chi_{n-r+1} \in \mathcal{H}_{n+1}$, and both sides of the equation to be proved are, for given $\zeta \in \Pi$, continuous linear forms on $\mathcal{H}_{n-r+1}$, of which $\mathcal{H}_{n+1} \cap \mathcal{H}_{n-r+1}$ is a dense subspace.

This concludes the proof of Theorem 3.2.6, also showing the validity of (3.2.58) in the case when $h \in (\mathcal{S}'_{\mathrm{weak}}(\mathbb{R}^2))^{\uparrow}$. $\square$

Remark 3.2.1. (i) It is also possible to obtain χ_{m+1} (or h_m) directly (i.e., not relying on a preliminary knowledge of the functions χ_{m+1-2r} with $r \geq 1$) as a linear combination of the functions $T_{m+1}^{k+m+1,k}$ for $[\frac{m-1}{2}]$ distinct nonnegative values of k: this is a consequence of (3.2.58) together with Lemma 3.2.3. If $\lambda_0, \ldots, \lambda_{[\frac{m-1}{2}]}$, depending on $(m; k_0, \ldots, k_{[\frac{m-1}{2}]})$, are chosen such that

$$\begin{pmatrix} 1 \\ 0 \\ \cdots \\ 0 \end{pmatrix} = \sum_{v=0}^{[\frac{m-1}{2}]} \lambda_v \begin{pmatrix} F_m^{k_v+m+1,k_v} \\ F_{m-2}^{k_v+m+1,k_v} \\ \cdots \\ F_{m-2\mu}^{k_v+m+1,k_v} \end{pmatrix}, \quad (3.2.62)$$

one has

$$\chi_{m+1}(\zeta) = \sum_{v=0}^{[\frac{m-1}{2}]} \lambda_v\, T_{m+1}^{k_v+m+1,k_v}(\zeta). \quad (3.2.63)$$

(ii) If $h \in (\mathcal{S}'_{\mathrm{weak}}(\mathbb{R}^2))^{\uparrow}$ is actually, with the same notation as before, a sum $\sum_{m \geq m_0} h_m$ for some $m_0 \geq 1$, one must complete the statement of Theorem 3.2.6 by the additional fact that, as seen from (3.2.58), $(\phi_\zeta^j | [X_1, [X_2, \ldots [X_\ell, B] \ldots]] \phi_\zeta^k)$ is zero unless $j - k \geq \ell + m_0 + 1$. We shall state Theorem 3.2.8 (the converse of Theorem 3.2.6) in a way including this extra piece of information: it will help in the proof by induction of that theorem.

(iii) In particular, $(\phi_\zeta^j \,|\, \mathrm{Op}^{\mathrm{asc}}(h)\,\phi_\zeta^k) = 0$ for every $h \in (S'_{\mathrm{weak}}(\mathbb{R}^2))^\uparrow$ and every $\zeta \in \Pi$ unless $j - k \geq 2$. A proper understanding of this fact requires that, together with the symbolic calculus $\mathrm{Op}^{\mathrm{asc}}$, one consider a "conjugate" one, substituting in (3.1.11) the operator $A_z^* = \pi^{\frac{1}{2}}(Q - zP)$ for $A_z = \pi^{\frac{1}{2}}(Q - \bar{z}P)$. In the conjugate calculus, the above condition changes to $k - j \geq 2$. The formula

$$\mathrm{Ana}\left(\begin{pmatrix} i & 0 \\ 0 & -i \end{pmatrix}\right) A_z \, \mathrm{Ana}\left(\begin{pmatrix} -i & 0 \\ 0 & i \end{pmatrix}\right) = -i A_{-\bar{z}}^*, \tag{3.2.64}$$

involving the transformation $u \mapsto u_i$, $u_i(x) = u(ix)$, introduced just after Theorem 2.2.3, explains the link between the two calculi. As will be seen in Sect. 3.4, a composition formula exists for the first (hence for each) of the two symbolic calculi. But the two families of operators obtained do not intersect, and taking the adjoint of an operator demands that one should move from one calculus to the other. In particular, the identity operator, for instance, does not belong to any of them: one can also remark that the existence *and uniqueness* of a symbol for the identity operator would be incompatible, anyway, with any of the two requirements of covariance. In view of (3.2.51) again, one may regard the alternative pseudodifferential analysis as an *ascending* calculus, its conjugate as a *descending* one; both adjectives have to be taken in a strict sense. It may be advisable to use $(S'_{\mathrm{weak}}(\mathbb{R}^2))^\downarrow$, rather than $(S'_{\mathrm{weak}}(\mathbb{R}^2))^\uparrow$, as a space of symbols for the descending calculus.

Before we can state and prove a complete converse to the part of Theorem 3.2.6 dealing with symbols in $(S_{\mathrm{weak}}(\mathbb{R}^2))^\uparrow$, we must show that the associated operators satisfy a certain condition, which is the object of Lemma 3.2.7. Though it may look obvious, it is not, since Theorem 3.2.6 does not yield yet any information about scalar products such as $(\phi_\zeta^j \,|\, B\phi_{\zeta'}^k)$ for $\zeta' \neq \zeta$: this will be repaired in Sect. 3.3, under stronger assumptions relative to the symbol h.

Lemma 3.2.7. *With the same notation as in the theorem that precedes, assume that* $h = \sum_{m \geq 1} h_m \in (S_{\mathrm{weak}}(\mathbb{R}^2))^\uparrow$. *Then, the identity*

$$\frac{\partial}{\partial \zeta}\left(\phi_\zeta^j \,\middle|\, [X_1, [X_2, \ldots [X_\ell, B]\ldots]]\,\phi_\zeta^k\right)$$

$$= \left(\frac{\partial}{\partial \bar{\zeta}}\,\phi_\zeta^j \,\middle|\, [X_1, [X_2, \ldots [X_\ell, B]\ldots]]\,\phi_\zeta^k\right) + \left(\phi_\zeta^j \,\middle|\, [X_1, [X_2, \ldots [X_\ell, B]\ldots]]\,\frac{\partial}{\partial \zeta}\,\phi_\zeta^k\right) \tag{3.2.65}$$

holds.

Proof. First note, thanks to Theorem 3.2.6 and Lemma 2.2.9, that both sides of the equation involve only the h_m's with $m \leq j - k - \ell + 1$. It is immediate, because of Proposition 3.1.6, that the proof of the statement can be reduced to the case when $\ell = 0$, i.e., when no commutator is present. Using (2.2.53), to the effect that

$$(\phi_\zeta^j | A_z^{-m-1} \phi_\zeta^k) = \sum_{\ell \in \mathbb{Z}} (\phi_\zeta^j | \phi_z^\ell)(A_z^{*-m-1} \phi_z^\ell | \phi_\zeta^k), \tag{3.2.66}$$

it is easy, with the help of estimates such as (2.2.44), to justify the equation

$$\frac{\partial}{\partial \zeta}(\phi_\zeta^j | A_z^{-m-1} \phi_\zeta^k) = (\frac{\partial}{\partial \bar{\zeta}} \phi_\zeta^j | A_z^{-m-1} \phi_\zeta^k) + (\phi_\zeta^j | A_z^{-m-1} \frac{\partial}{\partial \zeta} \phi_\zeta^k). \tag{3.2.67}$$

That this equation implies (3.2.65) can then be seen in a way paralleling the proof of Lemma 3.2.4, after we have made the left-hand side of (3.2.67) explicit as

$$C_m^{j,k} (\operatorname{Im} \zeta)^{\frac{m-1}{2}} (\bar{z} - \bar{\zeta})^{\frac{-m+1+j-k}{2}} \times$$
$$\left[\frac{m+1+j-k}{4i} (\bar{z} - \zeta)^{\frac{-m-3-j+k}{2}} + \frac{k-j}{4i} (\bar{z} - \zeta)^{\frac{-m-1-j+k}{2}} \right]. \tag{3.2.68}$$

Incidentally, one could also make the right-hand side of (3.2.67) explicit, making use of the relations (2.2.36): equating the two sides would lead to the equation (we shall not need it)

$$(m+1+j-k)C_m^{j,k} = \gamma_j^* \gamma_{j+1}^* C_m^{j+2,k} - \gamma_k \gamma_{k-1} C_m^{j,k-2}, \tag{3.2.69}$$

not a trivial consequence of Lemma 3.2.2. □

Together with Theorem 3.2.6 and Remark 3.2.1(ii), the following is an intrinsic characterization of a class of operators from the ascending pseudodifferential calculus which will be found helpful in Sect. 3.4, when dealing with the composition of operators. This characterization is not unlike Beals' characterization [2] of classes of pseudodifferential operators of the usual kind, in that it depends on the consideration of iterated brackets of the operator under study with the operators Q and P. That the proof of Theorem 3.2.8 looks somewhat complicated is in the nature of things: operators B from the ascending pseudodifferential analysis are characterized by properties of scalar products such as $(\phi_\zeta^j | B \phi_\zeta^k)$ with the same ζ on both sides (cf. Remark 3.2.1(iii)). In usual pseudodifferential analysis, one is much less concerned [31] with the full sequence of eigenfunctions of just one harmonic oscillator than with the pair of ground states of two different (say, Heisenberg – translated) harmonic oscillators.

Theorem 3.2.8. *Let B be a collection of linear operators: $E_\zeta \to E_\zeta'$ and let $m_0 \geq 1$ be given. Assume that, for every sequence $X = (X_1, \dots, X_\ell)$ as in Theorem 3.2.6 and for every pair (j,k), a decomposition such as (3.2.51) holds, with $T_{X,s}^{j,k} \in \mathcal{H}_s$. Assume that the function on the left-hand side of (3.2.51) is zero unless $j - k \geq \ell + m_0 + 1$. Finally, assume that the operator B satisfies the condition (3.2.65) from Lemma 3.2.7. Then, there is a unique element $h = \sum_{m \geq m_0} h_m \in (\mathcal{S}_{\text{weak}}(\mathbb{R}^2))^{\uparrow}$, such that $\sum_{m \geq m_0} (\phi_\zeta^j | \operatorname{Op}_m^{\text{asc}}(h_m) \phi_\zeta^k) = (\phi_\zeta^j | B \phi_\zeta^k)$ for every $\zeta \in \Pi$ and every pair (j,k).*

Proof. The proof consists in constructing $h_{m_0} \in S_{m_0}(\mathbb{R}^2)$ such that the operator $B_1 := B - \mathrm{Op}^{\mathrm{asc}}(h_{m_0})$ satisfies the same assumptions as those relative to B, except for the change of m_0 to $m_0 + 1$.

We first show how to build h_{m_0} as an element of $L^2_{m_0}(\mathbb{R}^2)$, such that the identity

$$(\phi_\zeta^{k+m_0+1} \,|\, \mathrm{Op}_{m_0}^{\mathrm{asc}}(h_{m_0})\, \phi_\zeta^k) = (\phi_\zeta^{k+m_0+1} \,|\, B\phi_\zeta^k) \qquad (3.2.70)$$

should hold for every $\zeta \in \Pi$ and every pair (j,k). With the notation of (3.2.59), one has $j - k = m_0 + 1$ so that $m \le m_0$ and, using (3.2.58), one sees that one can have $m + 1 - 2r \ge m_0 + 1$ there only if $m = m_0$ and $r = 0$. Hence, there is no choice but to define h_{m_0} in such a way that

$$(\mathrm{Im}\,\zeta)^{\frac{-m_0-1}{2}} (\phi_\zeta^{k+m_0+1} \,|\, B\phi_\zeta^k) = \frac{\pi^{\frac{m_0+1}{2}}}{m_0!} F_{m_0}^{k+m_0+1,k}(\Theta_{m_0} h_{m_0})(\zeta). \qquad (3.2.71)$$

If we do this, the function $\Theta_{m_0} h_{m_0}$ so defined lies in $\mathcal{H}_{m_0+1}$ by assumption, which is just what is needed in order that h_{m_0} should be well defined as an element of $L^2_{m_0}(\mathbb{R}^2)$: however, we must show that this definition of $\Theta_{m_0} h_{m_0}$ does not depend on k, i.e., that

$$(\phi_\zeta^{k+m_0+2} \,|\, B\phi_\zeta^{k+1}) = \frac{F_{m_0}^{k+m_0+2,k+1}}{F_{m_0}^{k+m_0+1,k}} (\psi_\zeta^{k+m_0+1} \,|\, B\phi_\zeta^k). \qquad (3.2.72)$$

According to (2.2.34), the left-hand side can be written as

$$\frac{(\mathrm{Im}\,\zeta)^{-\frac{1}{2}}}{\gamma_{k+m_0+1}^*} (A_\zeta^* \phi_\zeta^{k+m_0+1} \,|\, B\phi_\zeta^{k+1}) = \frac{(\mathrm{Im}\,\zeta)^{-\frac{1}{2}}}{\gamma_{k+m_0+1}^*} (\phi_\zeta^{k+m_0+1} \,|\, A_\zeta B\phi_\zeta^{k+1}). \qquad (3.2.73)$$

One has $(\phi_\zeta^{k+m_0+1} \,|\, [A_\zeta, B]\, \phi_\zeta^{k+1}) = 0$ because A_ζ is a linear combination of Q and P and, from one assumption of the theorem, $(\phi_\zeta^j \,|\, [Q \text{ or } P, B]\, \phi_\zeta^{k+1})$ can be nonzero only if $j - k - 1 \ge m_0 + 2$. It follows, using (2.2.34) again, that

$$(\phi_\zeta^{k+m_0+2} \,|\, B\phi_\zeta^{k+1}) = \frac{\gamma_{k+1}}{\gamma_{k+m_0+1}^*} (\phi_\zeta^{k+m_0+1} \,|\, B\phi_\zeta^k), \qquad (3.2.74)$$

and all that has to be shown is that

$$\frac{F_{m_0}^{k+m_0+2,k+1}}{F_{m_0}^{k+m_0+1,k}} = \frac{\gamma_{k+1}}{\gamma_{k+m_0+1}^*} \qquad (3.2.75)$$

for every $k \in \mathbb{Z}$. Using (2.2.34) together with (3.2.2), one may check that, according to whether one has $k \ge 0$, $-m_0 - 1 \le k \le -1$ or $k \le -m_0 - 2$, both sides of this equation are equal to $k + \frac{1}{2}$, 1 or $(k + m_0 + \frac{3}{2})^{-1}$.

Next, we prove that the function h_{m_0} lies in $S_{m_0}(\mathbb{R}^2)$. To this effect, setting $X = (X_1, \dots, X_{\alpha+\beta})$ with $X_1 = \dots = X_\alpha = P$, $X_{\alpha+1} = \dots = X_{\alpha+\beta} = Q$, we shall prove that the term from the decomposition (3.2.51) of $(\phi_\zeta^{k+m_0+1+\alpha+\beta} \,|\, [(\operatorname{ad} P)^\alpha \,(\operatorname{ad} Q)^\beta \,(B)] \,\phi_\zeta^k)$ with the highest power of $\operatorname{Im}\zeta$ is given, for some constant λ, as

$$(\operatorname{Im}\zeta)^{\frac{m_0+1+\alpha+\beta}{2}} T_{X,m_0+1+\alpha+\beta}^{k+m_0+1+\alpha+\beta,k}(\zeta)$$

$$= \lambda\,(\operatorname{Im}\zeta)^{\frac{m_0+1+\alpha+\beta}{2}} \left(\frac{d}{d\zeta}\right)^\alpha (\zeta\frac{d}{d\zeta}+m_0+\beta)\dots(\zeta\frac{d}{d\zeta}+m_0+1)\, T_{m_0+1}^{k+m_0+1,k}(\zeta):$$
$$(3.2.76)$$

recall that $(\operatorname{ad} X)(Y) = [X,Y]$ for any two operators X and Y, and observe that the left-hand side has been taken from (3.2.51), an assumption of the theorem to be proved. Note that the operators $\operatorname{ad} P$ and $\operatorname{ad} Q$ commute in view of Jacobi's identity and Heisenberg's relation. The case when $\alpha = \beta = 0$ is part of the assumption: when $j - k = m_0 + 1$, the integer s on the right-hand side of (3.2.51) must be set to the value $m_0 + 1$. Setting

$$C = (\operatorname{ad} P)^\alpha \,(\operatorname{ad} Q)^\beta \,(B), \qquad (3.2.77)$$

we then have to see the effect on the scalar product to be analyzed of substituting $[Q,C]$ or $[P,C]$ for C.

In view of the equations

$$P = \frac{1}{2i\pi^{\frac{1}{2}}}\,(\operatorname{Im}\zeta)^{-1}\,(A_\zeta - A_\zeta^*), \qquad Q = \frac{1}{2i\pi^{\frac{1}{2}}}\frac{\zeta}{\operatorname{Im}\zeta}\,(A_\zeta - A_\zeta^*) + \frac{1}{\pi^{\frac{1}{2}}}A_\zeta^*, \quad (3.2.78)$$

a rewriting of (3.2.35), we shall first find an expression for $(\phi_\zeta^{k+m_0+2+\alpha+\beta} \,|\, [A_\zeta, C]\phi_\zeta^k)$ and for $(\phi_\zeta^{k+m_0+2+\alpha+\beta} \,|\, [A_\zeta^*, C]\,\phi_\zeta^k)$. The second one is easier to deal with. We must examine the scalar product

$$(\phi_\zeta^{k+m_0+2+\alpha+\beta} \,|\, [A_\zeta^*, C]\,\phi_\zeta^k) = (A_\zeta\,\phi_\zeta^{k+m_0+2+\alpha+\beta} \,|\, C\phi_\zeta^k) - (\phi_\zeta^{k+m_0+2+\alpha+\beta} \,|\, CA_\zeta^*\,\phi_\zeta^k)$$

$$= (\operatorname{Im}\zeta)^{\frac{1}{2}}\left[\gamma_{k+m_0+2+\alpha+\beta}\,(\phi_\zeta^{k+m_0+1+\alpha+\beta} \,|\, C\phi_\zeta^k) - \gamma_k^*\,(\phi_\zeta^{k+m_0+2+\alpha+\beta} \,|\, C\phi_\zeta^{k+1})\right],$$
$$(3.2.79)$$

a consequence of Lemma 2.2.9. Let us also note that the two terms are proportional and that this equation reduces to

$$(\phi_\zeta^{k+m_0+2+\alpha+\beta} \,|\, [A_\zeta^*, C]\,\phi_\zeta^k) = \frac{m_0+1+\alpha+\beta}{\gamma_{k+m_0+1+\alpha+\beta}^*}\,(\operatorname{Im}\zeta)^{\frac{1}{2}}\,(\phi_\zeta^{k+m_0+1+\alpha+\beta} \,|\, C\phi_\zeta^k).$$
$$(3.2.80)$$

To see this, first note that

$$(\phi_\zeta^{k+m_0+1+\alpha+\beta} \,|\, [A_\zeta, C]\,\phi_\zeta^{k+1}) = 0, \qquad (3.2.81)$$

because the difference $(k+m_0+1+\alpha+\beta)-(k+1)$ is strictly less than $\alpha+\beta+m_0+2$, which would be the minimum needed, according to the assumption made, to get a nonzero scalar product: indeed, $[A_\zeta, C]$ is a linear combination (with coefficients depending on $\zeta, \bar{\zeta}$) of expressions $[X_1, [X_2, \ldots [X_\ell, B] \ldots]]$ with $\ell = \alpha+\beta+1$. Then, using Lemma 2.2.9 again, one may write

$$
\begin{aligned}
\gamma^*_{k+m_0+1+\alpha+\beta} \, (\phi_\zeta^{k+m_0+2+\alpha+\beta} | C\phi_\zeta^{k+1}) &= (A_\zeta^* \, \phi_\zeta^{k+m_0+1+\alpha+\beta} | C\phi_\zeta^{k+1}) \\
&= (\phi_\zeta^{k+m_0+1+\alpha+\beta} | A_\zeta C\phi_\zeta^{k+1}) \\
&= (\phi_\zeta^{k+m_0+1+\alpha+\beta} | CA_\zeta \phi_\zeta^{k+1}) \\
&= \gamma_{k+1} \, (\phi_\zeta^{k+m_0+1+\alpha+\beta} | C\phi_\zeta^{k}).
\end{aligned} \tag{3.2.82}
$$

If, looking back at (3.2.79), one notes that

$$
\gamma^*_{k+m_0+2+\alpha+\beta} - \gamma^*_k \frac{\gamma_{k+1}}{\gamma^*_{k+m_0+1+\alpha+\beta}} = \frac{\gamma^*_{k+m_0+2+\alpha+\beta} \, \gamma^*_{k+m_0+1+\alpha+\beta} - \gamma_{k+1} \, \gamma^*_k}{\gamma^*_{k+m_0+1+\alpha+\beta}} \tag{3.2.83}
$$

and that $\gamma_{j+1} \gamma^*_j = j + \frac{1}{2}$ for every j as seen from a case-by-case verification, one ends up with (3.2.80).

We now consider what happens when we replace the operator C by $[A_\zeta, C]$. Exactly for the same reason as that which led to (3.2.81), with one more commutator involved, one has also

$$
(\phi_\zeta^{k+m_0+1+\alpha+\beta} | [A_\zeta, [A_\zeta, C]] \, \phi_\zeta^k) = 0, \tag{3.2.84}
$$

which will be used in the following sequence of equations:

$$
\begin{aligned}
\gamma^*_{k+m_0+1+\alpha+\beta} \, (\phi_\zeta^{k+m_0+2+\alpha+\beta} | [A_\zeta, C] \, \phi_\zeta^k) &= (A_\zeta^* \, \phi_\zeta^{k+m_0+1+\alpha+\beta} | [A_\zeta, C] \, \phi_\zeta^k) \\
&= (\phi_\zeta^{k+m_0+1+\alpha+\beta} | A_\zeta [A_\zeta, C] \, \phi_\zeta^k) \\
&= (\phi_\zeta^{k+m_0+1+\alpha+\beta} | [A_\zeta, C] A_\zeta \, \phi_\zeta^k) \\
&= \gamma_k \, (\phi_\zeta^{k+m_0+1+\alpha+\beta} | [A_\zeta, C] \, \phi_\zeta^{k-1}),
\end{aligned} \tag{3.2.85}
$$

an equation to be used later. The core of the proof consists in evaluating the scalar product

$$
\begin{aligned}
(\phi_\zeta^{k+m_0+2+\alpha+\beta} &| [A_\zeta, C] \, \phi_\zeta^k) \\
&= (A_\zeta^* \, \phi_\zeta^{k+m_0+2+\alpha+\beta} | C\phi_\zeta^k) - (\phi_\zeta^{k+m_0+2+\alpha+\beta} | CA_\zeta \, \phi_\zeta^k) \\
&= (\mathrm{Im}\,\zeta)^{\frac{1}{2}} \left[\gamma^*_{k+m_0+2+\alpha+\beta} \, (\phi_\zeta^{k+m_0+3+\alpha+\beta} | C\phi_\zeta^k) - \gamma_k \, (\phi_\zeta^{k+m_0+2+\alpha+\beta} | C\phi_\zeta^{k-1}) \right].
\end{aligned} \tag{3.2.86}
$$

Changing k to $k-1$, one also finds

$$(\phi_\zeta^{k+m_0+1+\alpha+\beta}\,|\,[A_\zeta,C]\,\phi_\zeta^{k-1}) = (\operatorname{Im}\zeta)^{\frac{1}{2}} \times$$
$$\left[\gamma_{k+m_0+1+\alpha+\beta}^*\,(\phi_\zeta^{k+m_0+2+\alpha+\beta}\,|\,C\phi_\zeta^{k-1}) - \gamma_{k-1}\,(\phi_\zeta^{k+m_0+1+\alpha+\beta}\,|\,C\phi_\zeta^{k-2})\right].$$
$$(3.2.87)$$

We now take a linear combination of the last two equations, with coefficients chosen so as to eliminate the terms $(\phi_\zeta^{k+m_0+2+\alpha+\beta}\,|\,C\phi_\zeta^{k-1})$ from the combination, finding

$$\gamma_{k+m_0+1+\alpha+\beta}^*\,(\phi_\zeta^{k+m_0+2+\alpha+\beta}\,|\,[A_\zeta,C]\,\phi_\zeta^k) + \gamma_k\,(\phi_\zeta^{k+m_0+1+\alpha+\beta}\,|\,[A_\zeta,C]\,\phi_\zeta^{k-1})$$
$$= (\operatorname{Im}\zeta)^{\frac{1}{2}} \times \left[\gamma_{k+m_0+1+\alpha+\beta}^*\,\gamma_{k+m_0+2+\alpha+\beta}^*\,(\phi_\zeta^{k+m_0+3+\alpha+\beta}\,|\,C\phi_\zeta^k)\right.$$
$$\left. - \gamma_k\,\gamma_{k-1}\,(\phi_\zeta^{k+m_0+1+\alpha+\beta}\,|\,C\phi_\zeta^{k-2})\right]. \quad (3.2.88)$$

According to (3.2.85), the left-hand side of this equation can also be written as

$$2\,\gamma_{k+m_0+1+\alpha+\beta}^*\,(\phi_\zeta^{k+m_0+2+\alpha+\beta}\,|\,[A_\zeta,C]\,\phi_\zeta^k), \quad (3.2.89)$$

while, with the help of (2.2.36), the right-hand side is also

$$(\operatorname{Im}\zeta)^{\frac{1}{2}}\left([k+m_0+1+\alpha+\beta-4i\,(\operatorname{Im}\zeta)\frac{\partial}{\partial\bar{\zeta}}]\,\phi_\zeta^{k+m_0+1+\alpha+\beta}\,|\,C\phi_\zeta^k\right)$$
$$- (\operatorname{Im}\zeta)^{\frac{1}{2}}\left(\phi_\zeta^{k+m_0+1+\alpha+\beta}\,|\,C\,[k-4i\,(\operatorname{Im}\zeta)\frac{\partial}{\partial\zeta}]\,\phi_\zeta^k\right). \quad (3.2.90)$$

We can now use (3.2.65) to transform this to

$$(\operatorname{Im}\zeta)^{\frac{1}{2}}\left[m_0+1+\alpha+\beta+4i\,(\operatorname{Im}\zeta)\frac{\partial}{\partial\zeta}\right](\phi_\zeta^{k+m_0+1+\alpha+\beta}\,|\,C\phi_\zeta^k). \quad (3.2.91)$$

Summing up, with C as given in (3.2.77), we have on one hand (3.2.80), on the other hand the equation

$$(\phi_\zeta^{k+m_0+2+\alpha+\beta}\,|\,[A_\zeta,C]\,\phi_\zeta^k) = \frac{1}{2\,\gamma_{k+m_0+1+\alpha+\beta}^*} \times$$
$$(\operatorname{Im}\zeta)^{\frac{1}{2}}\left[m_0+1+\alpha+\beta+4i\,(\operatorname{Im}\zeta)\frac{\partial}{\partial\zeta}\right](\phi_\zeta^{k+m_0+1+\alpha+\beta}\,|\,C\phi_\zeta^k). \quad (3.2.92)$$

Let $X_0 = P$ or Q and set $Y = (X_0, X_1, \ldots, X_{\alpha+\beta})$. Assume that, for some pair (α,β), (3.2.76) is true, and let us compute the "main term" $(\operatorname{Im}\zeta)^{\frac{m_0+2+\alpha+\beta}{2}}$ $T_{Y,m_0+2+\alpha+\beta}^{k+m_0+2+\alpha+\beta,k}(\zeta)$ of $(\phi_\zeta^{k+m_0+2+\alpha+\beta}\,|\,[X_0,C]\,\phi_\zeta^k)$.

When $X_0 = P$, a look at (3.2.78) and (3.2.92) shows that this is also the main term of

$$\frac{1}{2i\pi^{\frac{1}{2}}} (\text{Im } \zeta)^{-1} \times \frac{2i(\text{Im } \zeta)^{\frac{3}{2}}}{\gamma_{k+m_0+1+\alpha+\beta}^*} \frac{\partial}{\partial \zeta} (\phi_\zeta^{k+m_0+1+\alpha+\beta} \,|\, C\phi_\zeta^k) : \tag{3.2.93}$$

finally, using (3.2.76), it coincides with

$$\frac{\pi^{-\frac{1}{2}}\lambda}{\gamma_{k+m_0+1+\alpha+\beta}^*} (\text{Im } \zeta)^{\frac{m_0+2+\alpha+\beta}{2}} \times$$

$$\left(\frac{d}{d\zeta}\right)^{\alpha+1} (\zeta\frac{d}{d\zeta} + m_0 + \beta) \dots (\zeta\frac{d}{d\zeta} + m_0 + 1) \, T_{m_0+1}^{k+m_0+1,k}(\zeta). \tag{3.2.94}$$

When $X_0 = Q$, the main term to be computed coincides with that of

$$\frac{1}{2i\pi^{\frac{1}{2}}} \frac{\zeta}{\text{Im } \zeta} (\phi_\zeta^{k+m_0+2+\alpha+\beta} \,|\, [A_\zeta, C]\phi_\zeta^k) + \frac{1}{\pi^{\frac{1}{2}}} (\phi_\zeta^{k+m_0+2+\alpha+\beta} \,|\, [A_\zeta^*, C]\phi_\zeta^k) :$$
$$\tag{3.2.95}$$

it is obtained from (3.2.92) and (3.2.80) as

$$(\text{Im } \zeta)^{\frac{m_0+2+\alpha+\beta}{2}} \times \frac{\pi^{-\frac{1}{2}}\lambda}{\gamma_{k+m_0+1+\alpha+\beta}^*} \left[\zeta\frac{d}{d\zeta} + (m_0+1+\alpha+\beta)\right]$$

$$(\frac{d}{d\zeta})^\alpha (\zeta\frac{d}{d\zeta} + m_0 + \beta) \dots (\zeta\frac{d}{d\zeta} + m_0 + 1) \, T_{m_0+1}^{k+m_0+1,k}(\zeta). \tag{3.2.96}$$

Since

$$(\zeta\frac{d}{d\zeta} + m_0 + 1 + \alpha + \beta)(\frac{d}{d\zeta})^\alpha = (\frac{d}{d\zeta})^\alpha (\zeta\frac{d}{d\zeta} + \beta + m_0 + 1), \tag{3.2.97}$$

(3.2.76) is proved by induction on $\alpha + \beta$. This equation, together with Lemma 3.1.4, proves that h_{m_0} remains in $L^2(\mathbb{R}^2)$ after having been applied any number of the operators $i\pi\bar{w}$ or $\frac{\partial}{\partial w}$: consequently, it lies in $S_{m_0}(\mathbb{R}^2)$.

At the same time, we have proved the identity

$$(\phi_\zeta^j \,|\, [X_1, [X_2, \dots [X_\ell, \text{Op}_{m_0}^{\text{asc}}(h_{m_0})]\dots]] \, \phi_\zeta^k) = (\phi_\zeta^j \,|\, [X_1, [X_2, \dots [X_\ell, B]\dots]] \, \phi_\zeta^k) \tag{3.2.98}$$

for every pair (j,k) such that $j - k \le m_0 + 1 + \ell$ because, though only the "main term" (i.e., that with the highest power of Im ζ) of each of the two sides has been made fully explicit, the calculations above have shown that each term, say on the right-hand side, can be obtained as the image of $(\phi_\zeta^{k+m_0+1} \,|\, B\phi_\zeta^k)$ under some differential operator. The same applies when $\text{Op}_{m_0}^{\text{asc}}(h_{m_0})$ takes the place of B, so that (3.2.98) appears as a consequence of (3.2.70).

This completes the proof of Theorem 3.2.8. $\square$

3.3 The Resolvent of the Lowering Operator

In usual analysis, the annihilation operator is the quintessential noninvertible operator, since it kills the ground state of the harmonic oscillator. The same goes for all operators $Q + iP - \omega$, $\omega \in \mathbb{C}$, as they are unitarily equivalent in $L^2(\mathbb{R})$ in view of the following formula: if $\omega = \xi + i\eta$ and if the Heisenberg translations are denoted as in (2.2.13), one has

$$Q + iP - \omega = \exp(2i\pi(\eta Q - \xi P))(Q + iP)\exp(-2i\pi(\eta Q - \xi P)). \quad (3.3.1)$$

In anaplectic analysis, as has already been recalled – but we shall need to give a new proof of the fact – the lowering operator is an automorphism of the basic space $\mathfrak{A}$: (3.3.1) is still valid, substituting pseudo-unitarity for unitarity. In calculations, however, one can also use instead the equation

$$Q + iP - \omega = \exp(-2i\pi\omega P)(Q + iP)\exp(2i\pi\omega P) \quad (3.3.2)$$

since (at the price of losing pseudo-unitarity) complex translations are possible in anaplectic analysis.

Proposition 3.3.1. *Let $f = (f_0, f_1, f_{i,0}, f_{i,1})$ be the $\mathbb{C}^4$-realization of some function $u \in \mathfrak{A}$. The components of the $\mathbb{C}^4$-realization $g = (g_0, g_1, g_{i,0}, g_{i,1})$ of the function $u_1 = (Q + iP)^{-1}u$ are given as follows:*

$$g_0(x) = 2\pi \left[\int_0^x e^{-\pi(x^2 - y^2)} f_1(y)\,dy - e^{-\pi x^2} \int_0^\infty e^{-\pi y^2} f_{i,1}(y)\,dy \right],$$

$$g_{i,0}(x) = -2\pi e^{\pi x^2} \int_x^\infty e^{-\pi y^2} f_{i,1}(y)\,dy,$$

$$g_1(x) = 2\pi \left[\int_0^x e^{-\pi(x^2 - y^2)} f_0(y)\,dy + e^{-\pi x^2} \int_0^\infty e^{-\pi y^2} f_{i,0}(y)\,dy \right],$$

$$g_{i,1}(x) = 2\pi e^{\pi x^2} \int_x^\infty e^{-\pi y^2} f_{i,0}(y)\,dy. \quad (3.3.3)$$

In particular,

$$u_1(x) = 2\pi \left[\int_0^x e^{-\pi(x^2 - y^2)} u(y)\,dy - e^{-\pi x^2} \int_0^\infty e^{-\pi y^2} f_{i,1}(y)\,dy \right]. \quad (3.3.4)$$

Proof. According to (2.2.6) and (2.2.7), the system that has to be solved is

$$(Q + iP)g_0 = f_1, \qquad (-Q + iP)g_{i,0} = f_{i,1},$$
$$(Q + iP)g_1 = f_0, \qquad (Q - iP)g_{i,1} = f_{i,0}. \quad (3.3.5)$$

The first equation yields

$$g_0(x) = 2\pi \left[Ce^{-\pi x^2} + \int_0^x e^{-\pi(x^2 - y^2)} f_1(y)\,dy \right]. \quad (3.3.6)$$

As the Gaussian function does not lie in $\mathfrak{A}$, only one constant C will do: it is to be determined by the condition that $g_{i,0}$, as defined by its relation (2.2.2) to g_0, must be nice in the sense of Definition 2.2.1. What it means is that one should have (in some sense, not splitting the sum that follows) $g_0(+i\infty) + ig_0(-i\infty) = 0$: this is a "boundary-value" problem of a very unusual kind, absolutely typical in anaplectic analysis. Now,

$$
\begin{aligned}
\frac{g_{i,0}(x)}{2\pi} &= C e^{\pi x^2} + \frac{1-i}{2} \left[\int_0^{ix} e^{\pi(x^2+z^2)} f_1(z)\,dz + i \int_0^{-ix} e^{\pi(x^2+z^2)} f_1(z)\,dz \right] \\
&= C e^{\pi x^2} + \int_0^x e^{\pi(x^2-y^2)} \frac{1+i}{2} \left[f_1(iy) - i f_1(-iy) \right] dy \\
&= C e^{\pi x^2} + e^{\pi x^2} \int_0^x e^{-\pi y^2} f_{i,1}(y)\,dy.
\end{aligned}
\tag{3.3.7}
$$

It is only in the case when $C = -\int_0^\infty e^{-\pi y^2} f_{i,1}(y)\,dy$ that this function can satisfy the proper estimate near $+\infty$: it then reduces to

$$
\frac{g_{i,0}(x)}{2\pi} = -e^{\pi x^2} \int_x^\infty e^{-\pi y^2} f_{i,1}(y)\,dy,
\tag{3.3.8}
$$

certainly a $O(e^{-\pi \varepsilon x^2})$ for some $\varepsilon > 0$ in view of the same estimate satisfied by the function $f_{i,1}$.

The computation of g_0 and $g_{i,0}$ is over, at least if the reader agrees to depend on the result, already mentioned, that $Q + iP$ is indeed an automorphism of $\mathfrak{A}$. However, our proof in [38, p. 163] was based on a totally different method; also, the reader may worry why the function g_0, as obtained in (3.3.6), has to be a $O(e^{-\pi \varepsilon x^2})$ near $+\infty$. Let us describe the argument briefly. With some small $\alpha > 0$ to be chosen later, one can transform the integral $k(x) = \int_0^x e^{-\pi(x^2-y^2)} f_1(y)\,dy$ (this is the only worrisome term) in two different ways, moving the contour of integration: the first new contour goes from 0 to $i\alpha x$ along the imaginary axis, then to αx along one-fourth of a circle centered at 0, finally from αx to x along the real line; the second contour is the complex conjugate of the first. We replace $k(x)$ by the linear combination of the integrals obtained, with coefficients $\frac{1-i}{2}$ and $\frac{1+i}{2}$, respectively. The net result is

$$
\begin{aligned}
k(x) &= \int_0^{\alpha x} e^{-\pi(x^2+y^2)} \frac{1+i}{2} \left(f_1(iy) - i f_1(-iy) \right) dy \\
&+ \frac{1-i}{2} \int_{-\frac{\pi}{2}}^{\frac{\pi}{2}} f_1(\alpha x e^{i\theta})\, \alpha x e^{i\theta}\, e^{-\pi x^2(1-\alpha^2 e^{2i\theta})} \left[\mathrm{char}_{[-\frac{\pi}{2},0]}(\theta) - \mathrm{char}_{[0,\frac{\pi}{2}]}(\theta) \right] d\theta \\
&+ \int_{\alpha x}^x e^{-\pi(x^2-y^2)} f_1(y)\,dy.
\end{aligned}
\tag{3.3.9}
$$

The first line is certainly a $O(e^{-\pi \varepsilon x^2})$ near $+\infty$, as seen when expressed in terms of the function $f_{i,1}$; so is, immediately, the last one. The integral in the middle can be majorized with the help of the estimate

$$|f_1(\alpha x e^{i\theta})| \leq C e^{\pi R \alpha^2 x^2} \tag{3.3.10}$$

together with $|e^{-\pi x^2(1-\alpha^2 e^{2i\theta})}| \leq e^{-\pi x^2(1-\alpha^2)}$: we get the desired estimate provided that $1 - \alpha^2 > R\alpha^2$.

The other two components of g are obtained in the same way. The expression of u_1 comes from the decomposition

$$u_1(x) = \frac{1}{2}[g_0(x) + g_0(-x) + g_1(x) - g_1(-x)], \tag{3.3.11}$$

a consequence of Definition 2.2.1. $\qquad\square$

Proposition 3.3.2. *If $g = (g_0, g_1, g_{i,0}, g_{i,1})$ is the $\mathbb{C}^4$-realization of the function $u_2 = A_z^{-2} u$, the $\mathbb{C}^4$-realization of u still being $f = (f_0, f_1, f_{i,0}, f_{i,1})$, one has*

$$g_0(x) = -\frac{4\pi}{z^2}\left[\int_0^x (x-y) e^{\frac{i\pi}{z}(x^2-y^2)} f_0(y)\,dy + \int_0^\infty (x-y) e^{\frac{i\pi}{z}(x^2+y^2)} f_{i,0}(y)\,dy\right],$$

$$g_{i,0}(x) = -\frac{4\pi}{z^2}\int_x^\infty (x-y) e^{-\frac{i\pi}{z}(x^2-y^2)} f_{i,0}(y)\,dy,$$

$$g_1(x) = -\frac{4\pi}{z^2}\left[\int_0^x (x-y) e^{\frac{i\pi}{z}(x^2-y^2)} f_1(y)\,dy - \int_0^\infty (x+y) e^{\frac{i\pi}{z}(x^2+y^2)} f_{i,1}(y)\,dy\right],$$

$$g_{i,1}(x) = -\frac{4\pi}{z^2}\int_x^\infty (x-y) e^{-\frac{i\pi}{z}(x^2-y^2)} f_{i,1}(y)\,dy. \tag{3.3.12}$$

Finally,

$$u_2(x) = -\frac{4\pi}{z^2}\left[\int_0^x (x-y) e^{\frac{i\pi}{z}(x^2-y^2)} u(y)\,dy - \int_0^\infty [y f_{i,0}(y) + x f_{i,1}(y)] e^{\frac{i\pi}{z}(x^2+y^2)}\,dy\right]. \tag{3.3.13}$$

Proof. Applying twice the formulas of Proposition 3.3.1, one obtains, in the case when $z = i$, the equations

$$g_0(x) = 4\pi\left[\int_0^x (x-y) e^{-\pi(x^2-y^2)} f_0(y)\,dy + e^{-\pi x^2}\int_0^\infty (x-y) e^{-\pi y^2} f_{i,0}(y)\,dy\right],$$

$$g_{i,0}(x) = 4\pi e^{\pi x^2}\int_x^\infty (x-y) e^{-\pi y^2} f_{i,0}(y)\,dy,$$

$$g_1(x) = 4\pi\left[\int_0^x (x-y) e^{-\pi(x^2-y^2)} f_1(y)\,dy - e^{-\pi x^2}\int_0^\infty (x+y) e^{-\pi y^2} f_{i,1}(y)\,dy\right],$$

$$g_{i,1}(x) = 4\pi e^{\pi x^2}\int_x^\infty (x-y) e^{-\pi y^2} f_{i,1}(y)\,dy. \tag{3.3.14}$$

To obtain the result for general z, we write

$$z = \frac{ai}{ci + a^{-1}} \quad \text{with} \quad \begin{pmatrix} a & 0 \\ c & a^{-1} \end{pmatrix} = \begin{pmatrix} \dfrac{|z|}{(\operatorname{Im} z)^{\frac{1}{2}}} & 0 \\ \dfrac{\operatorname{Re} z}{(\operatorname{Im} z)^{\frac{1}{2}}|z|} & \dfrac{(\operatorname{Im} z)^{\frac{1}{2}}}{|z|} \end{pmatrix}. \tag{3.3.15}$$

Then, from Proposition 2.2.8,

$$A_z = \frac{i\pi^{\frac{1}{2}}(\operatorname{Im} z)^{\frac{1}{2}}|z|}{z} \operatorname{Ana}\left(\begin{pmatrix} a & 0 \\ c & a^{-1} \end{pmatrix}\right)(Q+iP)\operatorname{Ana}\left(\begin{pmatrix} a^{-1} & 0 \\ -c & a \end{pmatrix}\right): \qquad (3.3.16)$$

also,

$$\left(\operatorname{Ana}\left(\begin{pmatrix} a & 0 \\ c & a^{-1} \end{pmatrix}\right)u_2\right)(x) = a^{-\frac{1}{2}}u_2(a^{-1}x)\,e^{i\pi ca^{-1}x^2},$$

$$\left(\operatorname{Ana}\left(\begin{pmatrix} a^{-1} & 0 \\ -c & a \end{pmatrix}\right)u\right)(y) = a^{\frac{1}{2}}u(ay)\,e^{-i\pi cay^2}. \qquad (3.3.17)$$

One may look at (2.2.24) and (2.2.25) to recall how such transformations act on the components of the $\mathbb{C}^4$-realizations of the functions under study: do not forget the sign change in the exponent when dealing with a component such as $f_{i,0}$. One then obtains

$$g_0(x) = -\frac{4\pi z}{(\operatorname{Im} z)\bar{z}} \times \left[e^{i\pi ca^{-1}x^2} \int_0^{a^{-1}x} (a^{-1}x-y)\,e^{-\pi(a^{-2}x^2-y^2)}\,f_0(ay)\,e^{-i\pi cay^2}\,dy \right.$$

$$\left. + e^{i\pi ca^{-1}x^2}\,e^{-\pi a^{-2}x^2} \int_0^{\infty} (a^{-1}x-y)\,e^{-\pi y^2}\,f_{i,0}(ay)\,e^{i\pi cay^2}\,dy \right]. \qquad (3.3.18)$$

To complete the calculation of $g_0(x)$, it suffices to carry an obvious change of variable, noting at the same time that $a^{-2}-ca^{-1}i = -\frac{i}{\bar{z}}$ and that $\frac{z}{(\operatorname{Im} z)\bar{z}}a^{-2} = \frac{1}{\bar{z}^2}$. The other components are obtained in the same way, still exercising much care with the signs of exponents.

To obtain (3.3.13), we observe that the even part of the function $x \mapsto \int_0^x(x-y)\,e^{\frac{i\pi}{\bar{z}}(x^2-y^2)}\,f_0(y)\,dy$ can be obtained by replacing, under the integral sign, the function $f_0(y)$ by its even part $\frac{1}{2}(f_0(y)+f_0(-y)) = u_{\text{even}}(y)$. A similar treatment goes for the odd part of the first integral in the expression for $g_1(x)$. $\qquad \square$

We now compute an operator such as (3.1.11)

$$\operatorname{Op}_1^{\text{asc}}(h) = \frac{1}{4}\int_{\Pi}(\Theta_1 h)(z)A_z^{-2}(\operatorname{Im} z)^2\,d\mu(z). \qquad (3.3.19)$$

On $\mathbb{R}^2$ interpreted as a phase space, one sometimes uses the variables (q,p) instead of the more traditional (x,ξ), a convention that we adopt here since x has been overused.

Theorem 3.3.3. Let $h \in S_1^A(\mathbb{R}^2)$ be given by the equation

$$h(q,p) = k(q^2+p^2)(q-ip). \qquad (3.3.20)$$

For $u \in \mathfrak{A}$ with the $\mathbb{C}^4$-realization $f = (f_0, f_1, f_{i,0}, f_{i,1})$, one has the identity

$$(\mathrm{Op}_1^{\mathrm{asc}}(h)\,u)(x)$$
$$= -\pi\left[\int_0^x (x-y)\,k(x^2-y^2)\,u(y)\,dy - \int_0^\infty k(x^2+y^2)\,[y\,f_{i,0}(y)+x\,f_{i,1}(y)]\,dy\right].$$
(3.3.21)

Proof. First note that

$$(\Theta_1 h)(z) = z^{-2}\int_{\mathbb{R}^2}(q+ip)\,e^{-\frac{i\pi}{z}(q^2+p^2)}\,h(q,p)\,dq\,dp$$
$$= \pi z^{-2}\int_0^\infty \rho\,k(\rho)\,e^{-\frac{i\pi\rho}{z}}\,d\rho.$$
(3.3.22)

With the help of (3.3.19) and (3.3.13), one can give an integral expression for $(\mathrm{Op}_1^{\mathrm{asc}}(h)\,u)(x)$: for simplicity, we immediately perform the change of variable $z\mapsto -z^{-1}$, which preserves $d\mu(z)$ and changes $z^{-2}(\bar{z})^{-2}(\mathrm{Im}\,z)^2$ to $(\mathrm{Im}\,z)^2$, getting as a result

$$(\mathrm{Op}_1^{\mathrm{asc}}(h)\,u)(x) = -\pi^2\int_\Pi (\mathrm{Im}\,z)^2\,d\mu(z)\int_0^\infty \rho\,k(\rho)\,e^{i\pi\rho z}\,d\rho$$
$$\left[\int_0^x (x-y)\,e^{-i\pi(x^2-y^2)\bar{z}}\,u(y)\,dy - \int_0^\infty [y\,f_{i,0}(y)+x\,f_{i,1}(y)]\,e^{-i\pi(x^2+y^2)\bar{z}}\,dy\right].$$
(3.3.23)

Hence,

$$(\mathrm{Op}_1^{\mathrm{asc}}(h)\,u)(x) = F_1(x) + F_2(x)$$
(3.3.24)

with

$$F_1(x) = -\pi^2\int_0^x (x-y)\,u(y)\,dy\int_0^\infty e^{-\pi(x^2-y^2)\eta}\,d\eta$$
$$\int_{-\infty}^\infty e^{-i\pi(x^2-y^2)\xi}\,d\xi\int_0^\infty \rho\,k(\rho)\,e^{\pi\rho(i\xi-\eta)}\,d\rho \quad (3.3.25)$$

and

$$F_2(x) = \pi^2\int_0^\infty [y\,f_{i,0}(y)+x\,f_{i,1}(y)]\,dy\int_0^\infty e^{-\pi(x^2+y^2)\eta}\,d\eta$$
$$\int_{-\infty}^\infty e^{-i\pi(x^2+y^2)\xi}\,d\xi\int_0^\infty \rho\,k(\rho)\,e^{\pi\rho(i\xi-\eta)}\,d\rho. \quad (3.3.26)$$

Setting $(x^2-y^2)_+ = (x^2-y^2)\,\mathrm{char}\,(x^2-y^2>0)$, one may reduce the first term as

$$F_1(x) = -2\pi^2\int_0^\infty (x-y)\,u(y)\,dy\int_0^\infty e^{-\pi(x^2-y^2)\eta}\,k(x^2-y^2)\,(x^2-y^2)_+\,d\eta$$
$$= -\pi\int_0^x (x-y)\,k(x^2-y^2)\,u(y)\,dy: \qquad (3.3.27)$$

in a similar way,

$$F_2(x) = \pi \int_0^\infty k(x^2 + y^2) \left[y f_{i,0}(y) + x f_{i,1}(y) \right] dy. \qquad (3.3.28)$$

$\square$

To obtain the other terms from the decomposition of $\mathrm{Op}^{\mathrm{asc}}(h)$, we would need to find a closed expression of A_z^{-m-1} for $m = 1, 2, \ldots$: we only know it, yet, for $m = 1$. Using (3.1.7) in the form

$$(\pi^{\frac{1}{2}} \omega - A_z)^{-2} = \sum_{m \geq 1} m \, \omega^{m-1} \, \pi^{\frac{m-1}{2}} A_z^{-m-1}, \qquad (3.3.29)$$

we obtain, for $m \geq 1$,

$$A_z^{-m-1} = \frac{\pi^{\frac{1-m}{2}}}{m!} \frac{\partial^{m-1}}{\partial \omega^{m-1}} \Bigg|_{\omega=0} (A_z - \pi^{\frac{1}{2}} \omega)^{-2}. \qquad (3.3.30)$$

Pushing the (elementary) calculations to the end, one obtains an explicit, but much too complicated formula. For most purposes, it is better to rely on Proposition 3.1.6, which makes it possible to reduce the study of $\mathrm{Op}_m^{\mathrm{asc}}$ to that of $\mathrm{Op}_1^{\mathrm{asc}}$.

We now wish to find sufficient conditions regarding a symbol $h \in S_1^A(\mathbb{R}^2)$ in order that the associated operator acts as an endomorphism of $\mathfrak{A}$. It is thus necessary to examine the four components of the function $\mathrm{Op}_1^{\mathrm{asc}}(h) u$: to this effect, we need to look at the components of $A_z^{-2} u$. In a way similar to the proof of Proposition 3.3.2, only replacing the right-hand side of (3.3.13) by that of the first line of (3.3.12), we obtain the following. Under the assumptions of Proposition 3.3.2, the first component g_0 of the $\mathbb{C}^4$-realization of $\mathrm{Op}_1^{\mathrm{asc}}(h) u$ is

$$g_0(x)$$
$$= -\pi \left[\int_0^x (x-y) k(x^2 - y^2) f_0(y) \, dy + \int_0^\infty (x-y) k(x^2 + y^2) f_{i,0}(y) \, dy \right]. \qquad (3.3.31)$$

Similarly,

$$g_1(x)$$
$$= -\pi \left[\int_0^x (x-y) k(x^2 - y^2) f_1(y) \, dy - \int_0^\infty (x-y) k(x^2 + y^2) f_{i,1}(y) \, dy \right] \qquad (3.3.32)$$

and

$$g_{i,0}(x) = -\pi \int_x^\infty (x-y) k(y^2 - x^2) f_{i,0}(y) \, dy,$$
$$g_{i,1}(x) = -\pi \int_x^\infty (x-y) k(y^2 - x^2) f_{i,1}(y) \, dy. \qquad (3.3.33)$$

We have to demand more about symbols, to wit a stronger estimate when (q, p) is close from $\mathbb{R}^2$ in a certain sense. We shall say that an entire function $h = h(q, p)$

lies in the space $\mathcal{G}(\mathbb{R}^2)$ if there exist positive constants ε, θ_0 and C such that

$$\left| h \left(\operatorname{Re} \left((q+ip)\, e^{i\theta} \right), \operatorname{Im} \left((q+ip)\, e^{i\theta} \right) \right) \right| \leq C e^{-\pi \varepsilon (q^2 + p^2)} \qquad (3.3.34)$$

whenever q, p, and θ are real and $|\theta| \leq \theta_0$. With the help of Cauchy's formula, it is clear that $\mathcal{G}(\mathbb{R}^2) \subset \mathcal{S}^A(\mathbb{R}^2)$. Also, $\mathcal{G}(\mathbb{R}^2)$ is invariant under rotations, which makes it possible to define the spaces $\mathcal{G}_m(\mathbb{R}^2)$ for $m \in \mathbb{Z}$. Finally, this space is invariant under the transformations $T_{\alpha,\beta}$ introduced in Definition 3.1.1 as well as under two of the three transformations that occur in (3.1.4). But it is not invariant under the two-dimensional Fourier transformation: a fully invariant substitute for $\mathcal{G}_m(\mathbb{R}^2)$ could most easily be characterized with the help of Proposition 2.1.1, the demand on the Θ_m-transform of the symbol being that it should extend as a holomorphic function to some neighborhood, in the Riemann sphere $\mathbb{C} \cup \{\infty\}$, of the closure of Π, having at ∞ a zero of order $\geq m+1$. But this is not required for the validity of the following theorem.

Theorem 3.3.4. *Let $h \in \mathcal{G}_m(\mathbb{R}^2)$ with $m \geq 1$. Then, the operator $\operatorname{Op}_m^{\mathrm{asc}}(h)$ is an endomorphism of the space $\mathfrak{A}$.*

Proof. Proposition 3.1.6 and the argument between (3.1.4) and (3.1.5) make it possible to reduce the problem to the case when $m = 1$. We may then use (3.3.31)–(3.3.33) to obtain the components of the $\mathbb{C}^4$-realization of $\operatorname{Op}_m^{\mathrm{asc}}(h)\, u$ in terms of those of u. The only nontrivial problem is to show that each of the four functions under consideration is a $O(e^{\pi \varepsilon x^2})$ as $x \to \infty$. We have solved such a kind of problem in the beginning of the present section, between (3.3.8) and (3.3.10): the sole novelty is that the function $e^{-\pi(x^2 - y^2)}$ has to be replaced by $k(x^2 - y^2)$. The same argument works again since, on one hand, $|k(x^2 + y^2)| \leq C e^{-\pi \varepsilon (x^2 + y^2)}$; on the other hand, for small $\alpha > 0$, one has

$$\left| \operatorname{Im} \left(x^2 (1 - \alpha^2 e^{2i\theta}) \right) \right| \leq \frac{\alpha^2}{1 - \alpha^2} \operatorname{Re} \left(x^2 (1 - \alpha^2 e^{2i\theta}) \right) \qquad (3.3.35)$$

so that, since $h \in \mathcal{G}(\mathbb{R}^2)$, one has again $|k(x^2 (1 - \alpha^2 e^{2i\theta}))| \leq C \exp(-\pi \varepsilon x^2)$ for some fixed ε: this makes it possible to conclude provided that $R \alpha^2 < \varepsilon$. $\qquad \square$

To prepare for the study of the composition of two operators, we need to show that an operator B with a symbol in $\mathcal{G}_m(\mathbb{R}^2)$ satisfies the condition (3.2.65) from Lemma 3.2.7, in a way which will make this property automatically valid for the composition of two such operators as well. This is easy, in view of the estimates just obtained. For a nice function f_0 (cf. Definition 2.2.1), and a pair (R, ε) of strictly positive numbers, set

$$\| f_0 \|_{R,\varepsilon} = \sup_{z \in \mathbb{C}} \left(e^{-\pi R |z|^2} |f_0(z)| \right) + \sup_{x>0} \left(e^{\pi \varepsilon x^2} |f_0(x)| \right); \qquad (3.3.36)$$

then, if $(f_0, f_1, f_{i,0}, f_{i,1})$ is the $\mathbb{C}^4$-realization of some function $u \in \mathfrak{A}$, set

$$\| \| u \| \|_{R,\varepsilon} = \| f_0 \|_{R,\varepsilon} + \| f_1 \|_{R,\varepsilon} + \| f_{i,0} \|_{R,\varepsilon} + \| f_{i,1} \|_{R,\varepsilon} : \qquad (3.3.37)$$

say that $u \in \mathfrak{A}_{R,\varepsilon}$ if $\| \| u \| \|_{R,\varepsilon} < \infty$. Given $u \in \mathfrak{A}$, there always exists such a pair (R, ε). What follows from the estimates that precede is that, if $h \in \mathcal{G}_m(\mathbb{R}^2)$, one can associate to any pair (R, ε) a pair (R', ε') such that $Bu: = \mathrm{Op}^{\mathrm{asc}}(h) u \in \mathfrak{A}_{R',\varepsilon'}$ if $u \in \mathfrak{A}_{R,\varepsilon}$. Such a property obviously transfers to the product of two such operators: we now show that it also makes (3.2.65) automatically valid. Since B is a linear endomorphism of $\mathfrak{A}$, we may this time consider pseudoscalar products $(\phi_\zeta^j | B \phi_{\zeta'}^k)$ without having to specialize to the case when $\zeta' = \zeta$. In view of Lemma 2.2.9, $\frac{\partial}{\partial \zeta} \phi_\zeta^j(x)$ can also be expressed as the image of ϕ_ζ^j under a second-order differential operator in the variable x, so that we just need estimates, in the nature of the finiteness of some (R', ε')-norm, for the derivatives of the components of the $\mathbb{C}^4$-realization of a function in $\mathfrak{A}_{R,\varepsilon}$. So far as the global estimate in the complex plane is concerned, Cauchy's integral formula on a circle with fixed radius will do. It can also work for the estimate as $x \to \infty$, with the help of the following application of the Phragmén–Lindelöf lemma, reproduced from [38, p. 5]: if $u \in \mathfrak{A}_{R,\varepsilon}$, there exist $C > 0$ and $\theta_0 > 0$ such that $|f_0(xe^{i\theta})| \le C e^{-\frac{\pi}{2} \varepsilon x^2}$ for $x > 0$ and $|\theta| \le \theta_0$, and the same goes for the other components of the $\mathbb{C}^4$-realization of u.

3.4 The Composition Formula

We study here the composition of two operators with symbols h^1 and h^2 in $\mathcal{G}_{m_1}(\mathbb{R}^2)$ and $\mathcal{G}_{m_2}(\mathbb{R}^2)$: we show that it has a symbol h (in the ascending calculus), to be denoted as $h = h^1 \# h^2$ and called the sharp product of h^1 and h^2 according to general definitions.

Lemma 3.4.1. *Let $h^1 \in \mathcal{G}_{m_1}(\mathbb{R}^2)$ and $h^2 \in \mathcal{G}_{m_2}(\mathbb{R}^2)$: let $\chi^1 = \Theta_{m_1} h^1$, $\chi^2 = \Theta_{m_2} h^2$ and set $B_1 = \mathrm{Op}_{m_1}^{\mathrm{asc}}(h^1)$, $B_2 = \mathrm{Op}_{m_2}^{\mathrm{asc}}(h^2)$. Recall that these spaces have been defined at the end of Sect. 3.3 and that the composition $B = B_1 B_2$ is well defined as an endomorphism of $\mathfrak{A}$. The operator B, when viewed as a collection of operators: $E_\zeta \to E'_\zeta$ (cf. Lemma 3.2.4), can be written as $B = \mathrm{Op}^{\mathrm{asc}}(h)$ for some (unique) symbol $h \in (\mathcal{S}_{\mathrm{weak}}(\mathbb{R}^2))^\intercal$. The isotypic components h_m of h are zero unless $m = m_1 + m_2 + 1 + 2p$ for some $p = 0, 1, \dots$: under this assumption, there exist coefficients $(\gamma_{p,q}^{m_1,m_2})_{0 \le q \le p}$, depending only on the integers indicated, such that*

$$(\Theta_m h_m)(\zeta) = \sum_{q=0}^{p} \gamma_{p,q}^{m_1,m_2} (\chi^1)^{(p-q)}(\zeta) (\chi^2)^{(q)}(\zeta), \qquad (3.4.1)$$

where the superscripts on the right-hand side mean that derivations with respect to ζ have to be carried, to the orders indicated.

Proof. We shall apply Theorem 3.2.8. That B satisfies the condition (3.2.65) has been proved at the very end of Sect. 3.3. We must first examine a pseudoscalar product such as

$$(\phi_\zeta^j \,|\, B_1 B_2 \,\phi_\zeta^k) = (B_1^* \,\phi_\zeta^j \,|\, B_2 \,\phi_\zeta^k) = \sum_{n \in \mathbb{Z}} c_n \,(\phi_\zeta^j \,|\, B_1 \,\phi_\zeta^n)\,(\phi_\zeta^n \,|\, B_2 \,\phi_\zeta^k), \qquad (3.4.2)$$

where the coefficients c_n have been defined and made explicit at the end of Sect. 2.2. According to Theorem 3.2.6,

$$(\phi_\zeta^j \,|\, B_1 \,\phi_\zeta^n) = \sum_{s_1 - 1 \in \mathfrak{S}(j-n-1)} (\mathrm{Im}\, w)^{\frac{s_1}{2}} \, T_{s_1}^{j,n}[h^1](\zeta),$$

$$(\phi_\zeta^n \,|\, B_2 \,\phi_\zeta^k) = \sum_{s_2 - 1 \in \mathfrak{S}(n-k-1)} (\mathrm{Im}\, w)^{\frac{s_2}{2}} \, T_{s_2}^{n,k}[h^2](w), \qquad (3.4.3)$$

with $T_{s_1}^{j,n}[h^1] \in \mathcal{H}_{s_1}$ and $T_{s_2}^{n,k}[h^2] \in \mathcal{H}_{s_2}$. Hence,

$$(\phi_\zeta^j \,|\, B \,\phi_\zeta^k) = \sum_s (\mathrm{Im}\,\zeta)^{\frac{s}{2}} \, T_s^{j,k}(\zeta) \qquad (3.4.4)$$

with

$$T_s^{j,k}(\zeta) = \sum_{s_1, s_2, n} c_n \, T_{s_1}^{j,n}[h^1](\zeta) \, T_{s_2}^{n,k}[h^2](\zeta), \qquad (3.4.5)$$

where the constraints on (s_1, s_2, n) are the following:

$$s_1 - 1 \in \mathfrak{S}(j-n-1), \quad s_2 - 1 \in \mathfrak{S}(n-k-1), \quad s_1 + s_2 = s, \quad n \in \mathbb{Z}, \quad (3.4.6)$$

to which we may add that

$$m_1 \in \mathfrak{S}(s_1 - 1), \qquad m_2 \in \mathfrak{S}(s_2 - 1), \qquad (3.4.7)$$

as is necessary, as observed from (3.2.58), for $T_{s_1}^{j,n}$ (resp. $T_{s_2}^{n,k}$) to be nonzero. The inequalities

$$m_1 \le s_1 - 1 \le j - n - 1 \qquad m_2 \le s_2 - 1 \le n - k - 1 \qquad (3.4.8)$$

for all nonzero terms imply $m_2 + k + 1 \le n \le j - m_1 - 1$ so that, in (3.4.5), we are dealing with a finite sum, not a series. Next, as a first step (we shall consider commutators with Q's and P's presently) toward applying Theorem 3.2.8, we must verify that $(\phi_\zeta^j \,|\, B \,\phi_\zeta^k) = 0$ unless $j - k \ge m_1 + m_2 + 2$, and that $T_s^{j,k}$ lies in $\mathcal{H}_s$. The first point is again a consequence of (3.4.8). The second one is the fact [34, p. 129] that the pointwise product of two functions, one in $\mathcal{H}_{s_1}$ and the other in $\mathcal{H}_{s_2}$, lies in $\mathcal{H}_{s_1 + s_2}$, a convexity estimate based on the Laplace representation (3.2.56).

It is immediate by induction that there are identities

$$[X_1, [X_2, \ldots, [X_\ell, B_1 B_2] \ldots]]$$
$$= \sum_{\ell_1 + \ell_2 = \ell} \beta_{\ell_1, \ell_2}^\ell \, [Y_1, [Y_2, \ldots, [Y_{\ell_1}, B_1] \ldots]] \, [Z_1, [Z_2, \ldots, [Z_{\ell_2}, B_2] \ldots]], \qquad (3.4.9)$$

in which each operator X, Y, or Z coincides with Q or P, and the X_r's are given arbitrarily. In view of Lemma 3.1.6, an operator such as $[Y_1, [Y_2, \ldots, [Y_{\ell_1}, B_1] \ldots]]$ has an isotypic symbol in $\mathcal{G}_{m_1+\ell_1}(\mathbb{R}^2)$, and something similar goes for B_2. It follows that the whole set of identities (3.2.51) making an application of Theorem 3.2.8 possible is satisfied, so that $B = B_1 B_2$ can be written as $\mathrm{Op}^{\mathrm{asc}}(h)$ for some $h = \sum_{m \geq m_1+m_2+1} h_m$, $h_m \in \mathcal{S}_m(\mathbb{R}^2)$.

What remains to be proved is that h_m can be nonzero only if m has the parity of $m_1 + m_2 + 1$, and that an identity such as (3.4.1) holds. To compute $\chi_{m+1} = \Theta_m h_m$ (do not confuse the superscript 1 of χ^1 or h^1 or the superscript 2 of χ^2 or h^2 with the subscript m, which specifies an isotypic component), we may use the special case when $p = 0$ of (3.2.58), to wit

$$T_{m+1}^{k+m+1,k}(\zeta) = \sum_{r=0}^{[\frac{m-1}{2}]} \frac{(2i)^r}{(m-r)!} \pi^{\frac{m-2r+1}{2}} F_{m-2r}^{k+m+1,k} \chi_{m+1-2r}^{(r)}(\zeta), \tag{3.4.10}$$

together with (3.4.5) of the left-hand side. When $j - k = s$, an integer ≥ 2 here denoted as $m + 1$, the inequalities (3.4.8), together with $s_1 + s_2 = m + 1$, force the condition

$$s_1 = j - n = m + 1 - n + k, \qquad s_2 = n - k, \tag{3.4.11}$$

and n is the only free parameter left in the sum under study: it is subject to the conditions $m_2 + k + 1 \leq n \leq j - 1 - m_1 = m - m_1 + k$. We have not yet used the conditions related to parity: $m_2 - s_2 + 1 = m_2 - n + k + 1$ is even, and so is $m_1 - s_1 + 1 = m_1 - m + n - k$; finally, the sum $m_1 + m_2 - m + 1$ is even too. We may thus set

$$n = m_2 + 1 + k + 2q, \qquad m = m_1 + m_2 + 1 + 2p, \tag{3.4.12}$$

and the inequalities relative to n become $0 \leq q \leq p$.

We may thus now assume that $m = m_1 + m_2 + 1 + 2p$ with $p = 0, 1, \ldots$ fixed and we sum up the information obtained in the equation

$$T_{m+1}^{k+m+1,k}[h](\zeta)$$
$$= \sum_{q=0}^{p} c_{m_2+1+k+2q} \, T_{m_1+2(p-q)+1}^{k+m+1,m_2+1+k+2q}[h^1](\zeta) \, T_{m_2+1+2q}^{m_2+1+k+2q,k}[h^2](\zeta). \tag{3.4.13}$$

The expansion (3.2.57) of $T_{m_2+1+2q}^{m_2+1+k+2q,k}[h^2](\zeta)$ of course reduces to one term, since h^2 reduces to $h_{m_2}^2$: the free index denoted as r in the equation just quoted would now have to be fixed to the value such that $(m_2 + 1 + 2q) - 2r = m_2 + 1$, i.e., to the value $r = q$. Hence,

$$T_{m_2+1+2q}^{m_2+1+k+2q,k}[h^2](\zeta) = \frac{(2i)^q}{(m_2+q)!} \pi^{\frac{m_2+1}{2}} F_{m_2}^{m_2+1+k+2q,k} (\chi^2)^{(q)}(\zeta). \tag{3.4.14}$$

In a similar way, one obtains

$$T^{k+m+1,m_2+1+k+2q}_{m_1+2(p-q)+1}[h^1](\zeta) = \frac{(2i)^{p-q}}{(m_1+p-q)!}\,\pi^{\frac{m_1+1}{2}}\,F^{k+m+1,m_2+1+k+2q}_{m_1}\,(\chi^1)^{(p-q)}(\zeta).$$

$$(3.4.15)$$

We thus have an expression on the right-hand side of (3.4.13) fully similar to the right-hand side of (3.4.1). To find a similar expression for the function $\chi_{m+1}(\zeta)$, we rely on (3.4.10): a proof by induction, or one based on Lemma 3.2.3 as in Remark 2.2.1(i), taking advantage of (3.4.10) for $[\frac{m-1}{2}]$ distinct nonnegative values of k, makes it possible to conclude.   □

What remains to be done is making the coefficients of the expression on the right-hand side of (3.4.1) explicit, an easy task, up to a point, in view of the covariance of the ascending pseudodifferential calculus.

Lemma 3.4.2. *Let the Rankin–Cohen brackets [6] be the bilinear operations defined by the equation (in which $m = m_1 + m_2 + 1 + 2p$)*

$$K^{m_1+1,m_2+1}_{m+1}(\chi^1,\chi^2) = \sum_{q=0}^{p}(-1)^q\binom{m_1+p}{q}\binom{m_2+p}{p-q}(\chi^1)^{(p-q)}(\chi^2)^{(q)}:$$

$$(3.4.16)$$

the input functions χ^1 and χ^2 could be arbitrary elements of the spaces $\mathcal{H}_{m_1+1}$ and $\mathcal{H}_{m_2+1}$, respectively. Under the assumptions of Lemma 3.4.1, one has for some constant $\Gamma^{m_1+1,m_2+1}_{m+1}$ the identity

$$(\Theta_m h_m)(w) = \Gamma^{m_1+1,m_2+1}_{m+1}\,K^{m_1+1,m_2+1}_{m+1}(\chi^1,\chi^2)(w). \qquad (3.4.17)$$

Proof. The statement of Lemma 3.4.1 is that the function $\Theta_{m_1+m_2+1+2p}h_{m_1+m_2+1+2p}$ is obtained from $\chi^1 = \Theta_{m_1}h^1$, $\chi^2 = \Theta_{m_2}h^2$ by means of a bilinear operator, which we shall temporarily denote as $\mathcal{L}^{m_1+1,m_2+1}_{m+1}$, of the form characterized by the equation

$$(\mathcal{L}^{m_1+1,m_2+1}_{m+1}(\chi^1,\chi^2))(\zeta) = \sum_{q=0}^{p}\gamma_q\,(\chi^1)^{(p-q)}(\zeta)\,(\chi^2)^{(q)}(\zeta): \qquad (3.4.18)$$

fixing m_1, m_2, and p has made it possible to simplify notation. One has the identity

$$\mathcal{D}_{m_1+m_2+2+2p}(g)\,\mathcal{L}^{m_1+1,m_2+1}_{m_1+m_2+2+2p}(\chi^1,\chi^2) = \mathcal{L}^{m_1+1,m_2+1}_{m_1+m_2+2+2p}(\mathcal{D}_{m_1+1}(g)\chi^1,\mathcal{D}_{m_2+1}(g)\chi^2)$$

$$(3.4.19)$$

for every $g \in SL(2,\mathbb{R})$. Indeed, since the representation Met$^{(2)}$ of $SL(2,\mathbb{R})$ reduces on isotypic components of $(L^2(\mathbb{R}^2))^\uparrow$ to a set of representations, each of which transfers to $\mathcal{D}_{m+1}$ under Θ_m, this latter identity is a consequence of the covariance (3.1.29) of the ascending symbolic calculus under the anaplectic representation. Since the bilinear operation under study is given by a differential operator, it must remain true if one substitutes for χ_1 and χ_2 arbitrary holomorphic functions in Π, extending the three representations involved in the natural way. Take $\chi^1(\zeta) = \zeta^{-m_1-1}$, $\chi^2(\zeta) = 1$: then, $(\mathcal{L}_{m_1+m_2+2+2p}^{m_1+1,m_2+1}(\chi_1,\chi_2))(\zeta) = (-1)^p \frac{(m_1+p)!}{m_1!} \gamma_0 \zeta^{-m_1-1-p}$, and, if $g = \begin{pmatrix} a & b \\ c & d \end{pmatrix}$,

$$(\mathcal{D}_{m_1+1}(g)\chi^1)(\zeta) = (d\zeta - b)^{-m_1-1},$$
$$(\mathcal{D}_{m_1+1}(g)\chi^2)(\zeta) = (-c\zeta + a)^{-m_2-1}. \tag{3.4.20}$$

The identity expressing the covariance expresses itself, in this case, as

$$(-1)^p \frac{(m_1+p)!}{m_1!} \gamma_0 (-c\zeta + a)^{-m_1-m_2-2p-2} \left(\frac{d\zeta - b}{-c\zeta + a}\right)^{-m_1-1-p} = \sum_{q=0}^{p} \gamma_q$$
$$\frac{(m_1+p-q)!}{m_1!} \frac{(m_2+q)!}{m_2!} (-d)^{p-q} c^q (d\zeta - b)^{-m_1-1-p+q} (-c\zeta + a)^{-m_2-1-q}, \tag{3.4.21}$$

which reduces to the identity

$$\sum_{q=0}^{p} (-1)^{p-q} \gamma_q \frac{(m_1+p-q)!}{(m_1+p)!} \frac{(m_2+q)!}{m_2!} (cd\zeta - bc)^q (ad - cd\zeta)^{p-q} = 1 \tag{3.4.22}$$

or, introducing the indeterminate $\xi = cd\zeta - bc$, to the identity

$$\sum_{q=0}^{p} (-1)^{p-q} \gamma_q \frac{(m_1+p-q)!}{(m_1+p)!} \frac{(m_2+q)!}{m_2!} \xi^q (1-\xi)^{p-q} = 1. \tag{3.4.23}$$

Lemma 3.4.2 follows.   □

Next, we transfer the Rankin–Cohen operations under the Θ-transforms: note, going back to (3.3.34), that an immediate corollary of Lemma 3.4.3 is that the function $h_m \in S_m(\mathbb{R}^2)$ characterized by the identity (3.4.24) below lies in $\mathcal{G}_m(\mathbb{R}^2)$ if $h^1 \in \mathcal{G}_{m_1}(\mathbb{R}^2)$ and $h^2 \in \mathcal{G}_{m_2}(\mathbb{R}^2)$.

Lemma 3.4.3. *Assuming* $m_1 \geq 1$, $m_2 \geq 1$, $p \geq 0$, *let* $m = m_1 + m_2 + 1 + 2p$. *Let* $h^1 \in S_{m_1}(\mathbb{R}^2)$ *and* $h^2 \in S_{m_2}(\mathbb{R}^2)$, *and let* $h_m \in S_m(\mathbb{R}^2)$ *be characterized by the identity*

$$\Theta_m h_m = K_{m+1}^{m_1+1,m_2+1}(\Theta_{m_1} h^1, \Theta_{m_2} h^2). \tag{3.4.24}$$

Setting

$$h^1(x, \xi) = (x - i\xi)^{m_1} k^1(x^2 + \xi^2),$$
$$h^2(x, \xi) = (x - i\xi)^{m_2} k^2(x^2 + \xi^2),$$
$$h_m(x, \xi) = (x - i\xi)^{m} k_m(x^2 + \xi^2), \tag{3.4.25}$$

one has the identity

$$k_m(\rho) = \frac{(-i)^p \pi^{p+1}}{p!} \rho^{-p} \int_0^1 (1-t)^{m_1+p} t^{m_2+p}$$

$$\left(\frac{d}{dt}\right)^p \left[k^1(\rho(1-t))k^2(\rho t)\right] dt. \quad (3.4.26)$$

Proof. One has

$$\left(\mathcal{D}_{m+1}\left(\left(\begin{smallmatrix} 0 & 1 \\ -1 & 0 \end{smallmatrix}\right)\right)(\Theta_m h_m)\right)(z) = (-1)^{m+1} \int_{\mathbb{R}^2} (x+i\xi)^m e^{i\pi z(x^2+\xi^2)} h_m(x,\xi) \, dx d\xi$$

$$= (-1)^{m+1} \int_{\mathbb{R}^2} (x^2+\xi^2)^m e^{i\pi z(x^2+\xi^2)} k_m(x^2+\xi^2) \, dx d\xi$$

$$= (-1)^{m+1} \pi \int_0^\infty \rho^m e^{i\pi z \rho} k_m(\rho) \, d\rho. \quad (3.4.27)$$

Using analogous expressions in relation with the symbols h^1 and h^2, differentiating with respect to z under the integral sign, and taking advantage of the covariance of the Rankin–Cohen operation, one obtains the relation

$$\int_0^\infty \rho^m e^{i\pi z \rho} k_m(\rho) \, d\rho = i^p \pi^{p+1} \sum_{q=0}^p (-1)^q \binom{m_1+p}{q} \binom{m_2+p}{p-q}$$

$$\int_0^\infty \int_0^\infty \rho_1^{m_1+p-q} \rho_2^{m_2+q} e^{i\pi z(\rho_1+\rho_2)} k_1(\rho_1) k_2(\rho_2) \, d\rho_1 d\rho_2. \quad (3.4.28)$$

Noting the identity

$$\sum_{q=0}^p (-1)^q \binom{m_1+p}{q} \binom{m_2+p}{p-q} (1-t)^{m_1+p-q} t^{m_2+q}$$

$$= \frac{1}{p!} \left(\frac{d}{dt}\right)^p \left[(1-t)^{m_1+p} t^{m_2+p}\right], \quad (3.4.29)$$

and setting $\rho_1 = \rho t$, $\rho_2 = \rho(1-t)$ on the right-hand side of (3.4.28), we obtain

$$\rho^{m_1+m_2+1+2p} k_m(\rho) = \frac{i^p \pi^{p+1}}{p!} \rho^{m_1+m_2+1+p}$$

$$\times \int_0^1 k^1(\rho(1-t)) k^2(\rho t) \left(\frac{d}{dt}\right)^p \left[(1-t)^{m_1+p} t^{m_2+p}\right] dt, \quad (3.4.30)$$

so that the result follows after an integration by parts. $\qquad \square$

To determine the coefficient which occurs on the right-hand side of (3.4.17), we need to consider special, but important, operators: this is the object of next lemma.

Lemma 3.4.4. *Fixing $z \in \Pi$, set, for $m = 1,2,\dots$,*

$$h_m(x,\xi) = \bar{z}^{-m-1} (x-i\xi)^m \exp\left(\frac{i\pi}{\bar{z}} (x^2+\xi^2)\right), \quad (3.4.31)$$

so that $h_m \in \mathcal{S}_m(\mathbb{R}^2)$. One has

$$\chi_{m+1}(\zeta) := (\Theta_m h_m)(\zeta) = i^{m+1} m! \pi^{-m} (\zeta - \bar{z})^{-m-1} \tag{3.4.32}$$

and

$$\mathrm{Op}^{\mathrm{asc}}(h_m) = \pi^{\frac{1-m}{2}} 2^{-m-1} m! A_z^{-m-1}. \tag{3.4.33}$$

Proof. The first equation is just (3.2.31). From (3.2.51), (3.2.53), and (3.2.55), one has

$$(\phi_\zeta^j \mid \mathrm{Op}^{\mathrm{asc}}(h_m) \, \phi_\zeta^k) = \pi^{\frac{m+1}{2}} m! C_m^{j,k} \sum_{\substack{m+1 \le s \le j-k \\ s-m-1 \ \text{even}}}$$

$$(\mathrm{Im} \, \zeta)^{\frac{s}{2}} \binom{\frac{j-k-1-m}{2}}{\frac{s-m-1}{2}} (2i)^{\frac{s-m-1}{2}} \left(\frac{i}{2}\right)^{m+1} \frac{1}{(\frac{s+m-1}{2})!} \chi_{m+1}^{(\frac{s-m-1}{2})}(\zeta) \tag{3.4.34}$$

if $j - k - m - 1$ is an even nonnegative integer, while the result is zero otherwise. Also,

$$\chi_{m+1}^{(\frac{s-m-1}{2})}(\zeta) = \pi^{-m} (-i)^{m+1} \left(\frac{s+m-1}{2}\right)! (\bar{z} - \zeta)^{\frac{-m-1-s}{2}}. \tag{3.4.35}$$

On the other hand, from Lemma 3.2.1, one has if not in the zero case

$$(\phi_\zeta^j \mid A_z^{-m-1} \, \phi_\zeta^k) = C_m^{j,k} (\mathrm{Im} \, \zeta)^{\frac{m+1}{2}} (\bar{z} - \bar{\zeta})^{\frac{-m-1+j-k}{2}} (\bar{z} - \zeta)^{\frac{-m-1-j+k}{2}}. \tag{3.4.36}$$

After a binomial expansion and the change of index from $r = 0, \ldots, j - k - m - 1$ to $s = m + 1 + 2r$ (just as in the proof of Theorem 3.2.6), we obtain

$$(\phi_\zeta^j \mid A_z^{-m-1} \, \phi_\zeta^k) = C_m^{j,k} \times$$

$$\sum_{\substack{m+1 \le s \le j-k \\ s-m-1 \ \text{even}}} (2i)^{\frac{s-m-1}{2}} \binom{\frac{j-k-1-m}{2}}{\frac{s-m-1}{2}} (\mathrm{Im} \, \zeta)^{\frac{s}{2}} (\bar{z} - \zeta)^{\frac{-m-1-s}{2}}. \tag{3.4.37}$$

The lemma follows.   □

Theorem 3.4.5. *Let $h^1 \in \mathcal{G}_{m_1}(\mathbb{R}^2)$ and $h^2 \in \mathcal{G}_{m_2}(\mathbb{R}^2)$: let $\chi^1 = \Theta_{m_1} h^1$, $\chi^2 = \Theta_{m_2} h^2$ and set $B_1 = \mathrm{Op}_{m_1}^{\mathrm{asc}}(h^1)$, $B_2 = \mathrm{Op}_{m_2}^{\mathrm{asc}}(h^2)$. The composition $B_1 B_2$ has a symbol h in $\mathcal{G}_{m_1+m_2+1}(\mathbb{R}^2)$, the isotypic components of which are characterized by the equation*

$$(\Theta_{m_1+m_2+1+2p} \, h_{m_1+m_2+1+2p})(w) = \left(\frac{i}{\pi}\right)^p K_{m_1+m_2+2+2p}^{m_1+1, m_2+1}(\chi^1, \chi^2)(w). \tag{3.4.38}$$

Only isotypic components of the orders just indicated can occur in h.

Proof. To obtain the coefficient $\Gamma^{m_1+1,m_2+1}_{m_1+m_2+2+2p}$, as it would appear in (3.4.17), one can apply Lemma 3.4.2 in a particular case. We shall consider the two operators

$$B_1 = \left(\frac{\partial}{\partial \bar{z}}\right)^p [(Q - \bar{z}P)^{-m_1-1}](z = i), \qquad B_2 = (Q + iP)^{-m_2-1}. \qquad (3.4.39)$$

Note that the symbol h^1 (resp. h^2) of B_1 (resp. B_2) reduces to its m_1-isotypic (resp. m_2-isotypic) component. According to Lemma 3.4.4, one has

$$\chi^2(\zeta): = (\Theta_{m_2} h^2)(\zeta) = \left(\frac{\zeta + i}{2i}\right)^{-m_2-1}. \qquad (3.4.40)$$

Also, the (m_1-isotypic) symbol $h^{(z)}_{m_1}$ of $A_z^{-m_1-1}$ is characterized by the equation

$$(\Theta_{m_1} h^{(z)}_{m_1})(\zeta) = \left(\frac{\zeta - \bar{z}}{2i}\right)^{-m_2-1}: \qquad (3.4.41)$$

consequently, the Θ_{m_1}-transform of the (m_1-isotypic) symbol of B_1 is

$$\chi^1(\zeta): = (\Theta_{m_1} h^1)(\zeta) = (2i)^{m_1+1} \frac{(m_1 + p)!}{m_1!} (\zeta + i)^{-m_1-1-p}. \qquad (3.4.42)$$

It is an easy task to compute the Rankin–Cohen bracket

$$\mathcal{K}^{m_1+1,m_2+1}_{m_1+m_2+2+2p}(\chi^1, \chi^2)(\zeta) = (2i)^{m_1+m_2+2} \frac{(m_1 + p)!}{m_1!} \sum_{q=0}^{p}(-1)^q$$
$$\binom{m_1 + p}{q}\binom{m_2 + p}{p - q}\left(\frac{\partial}{\partial \zeta}\right)^{(p-q)}(\zeta + i)^{-m_1-1-p}\left(\frac{\partial}{\partial \zeta}\right)^{(q)}(\zeta + i)^{-m_2-1}: \qquad (3.4.43)$$

this reduces to

$$\mathcal{K}^{m_1+1,m_2+1}_{m_1+m_2+2+2p}(\chi^1, \chi^2)(\zeta)$$
$$= (2i)^{m_1+m_2+2} \frac{(m_1 + p)!\,(m_2 + p)!}{m_1!\,m_2!} (\zeta + i)^{-m_1-m_2-2-2p} \times a_p \qquad (3.4.44)$$

with

$$a_p = \left(\frac{\partial}{\partial x}\right)^p \left[\sum_{q=0}^{p}(-1)^{p-q}\frac{x^{m_1+2p-q}}{q!\,(p-q)!}\right](x = 1)$$
$$= \left(\frac{\partial}{\partial x}\right)^p \left[x^{m_1+p}\frac{(1-x)^p}{p!}\right](x = 1) = (-1)^p. \qquad (3.4.45)$$

On the other hand, starting from (3.2.17), one shows by induction that, for every $n \geq 1$ and $s = 0, 1, \ldots$,

$$
\begin{aligned}
D_s^n :&= \left(\frac{\partial}{\partial \bar{z}}\right)^s \bigg|_{z=i} (Q - \bar{z}P)^{-n-1} \\
&= \sum_{\ell=0}^{s} (4i\pi)^{-\ell} \frac{(n+s+\ell)!}{n!} \binom{s}{\ell} P^{s-\ell} (Q+iP)^{-n-1-s-\ell}.
\end{aligned}
\tag{3.4.46}
$$

In particular, it follows that

$$
B_1 B_2 = \sum_{j=0}^{p} (4i\pi)^{-j} \frac{(m_1 + p + j)!}{m_1!} \binom{p}{j} P^{p-j} (Q+iP)^{-m_1 - m_2 - 2 - p - j}. \tag{3.4.47}
$$

Let $m = m_1 + m_2 + 1$ and set, for $0 \leq r \leq p, \, 0 \leq j \leq p$,

$$
\begin{aligned}
X_r &= \frac{(4i\pi)^{-r} (m + 2r)!}{(p-r)!} D_{p-r}^{m+2r}, \\
Y_j &= \frac{(4i\pi)^{-j} (m + p + j)!}{(p-j)!} P^{p-j} (Q+iP)^{-m-1-p-j}.
\end{aligned}
\tag{3.4.48}
$$

Relations (3.4.46) reduce to

$$
X_r = \sum_{j=r}^{p} \frac{1}{(j-r)!} Y_j \qquad 0 \leq r \leq p, \tag{3.4.49}
$$

a triangular system solved as

$$
Y_j = \sum_{r=j}^{p} \frac{(-1)^{r-j}}{(r-j)!} X_r \qquad 0 \leq j \leq p. \tag{3.4.50}
$$

Then, one may rewrite (3.4.47) as

$$
\begin{aligned}
B_1 B_2 &= \frac{p!}{m_1!} \sum_{j=0}^{p} \frac{(m_1 + p + j)!}{j!(m+p+j)!} Y_j \\
&= \frac{p!}{m_1!} \sum_{j=0}^{p} \frac{(m_1 + p + j)!}{j!(m+p+j)!} \sum_{r=j}^{p} \frac{(-1)^{r-j}}{(r-j)!} (4i\pi)^{-r} \frac{(m+2r)!}{(p-r)!} D_{p-r}^{m+2r}. \tag{3.4.51}
\end{aligned}
$$

This equation is equivalent to the decomposition of the symbol h of $B_1 B_2$ into its isotypic components, which have the types $m_1 + m_2 + 1, m_1 + m_2 + 3, \ldots, m_1 + m_2 + 1 + 2p$. The coefficient of $D_0^{m+2p} = D_0^{m_1 + m_2 + 1 + 2p}$ there is

$$
(4i\pi)^{-p} \frac{p!(m+2p)!}{m_1!} \sum_{j=0}^{p} \frac{(-1)^{p-j}}{j!(p-j)!} \frac{(m_1 + p + j)!}{(m+p+j)!}. \tag{3.4.52}
$$

Since the $\Theta_{m_1+m_2+1+2p}$-transform of the $((m_1+m_2+1+2p)$-isotypic) symbol of $D_0^{m_1+m_2+1+2p}$ is (3.4.40), the function $\zeta \mapsto \left(\frac{\zeta+i}{2i}\right)^{-m_1-m_2-2-2p}$, one finds that the (m_1+m_2+1+2p)-isotypic component of h is given by the equation

$$(\Theta_{m_1+m_2+1+2p}\, h_{m_1+m_2+1+2p})(\zeta) = (4i\pi)^{-p}\, \frac{p!\,(m_1+m_2+1+2p)!}{m_1!}$$

$$\times \sum_{j=0}^{p} \frac{(-1)^{p-j}}{j!\,(p-j)!}\, \frac{(m_1+p+j)!}{(m_1+m_2+1+p+j)!} \times \left(\frac{\zeta+i}{2i}\right)^{-m_1-m_2-2-2p}. \qquad (3.4.53)$$

Comparing this expression with (3.4.44), one obtains that the coefficient which occurs in (3.4.17) is

$$\Gamma_{m_1+m_2+2+2p}^{m_1+1,m_2+1} = (\frac{i}{\pi})^p\, \frac{p!\,m_2!\,(m_1+m_2+1+2p)!}{(m_1+p)!\,(m_2+p)!}\, S_p(m_1,m_2) \qquad (3.4.54)$$

with

$$S_p(m_1,m_2) = \sum_{j=0}^{p} \frac{(-1)^j}{j!\,(p-j)!}\, \frac{(m_1+p+j)!}{(m_1+m_2+1+p+j)!}. \qquad (3.4.55)$$

This sum is the value at $x=1$ of the polynomial the (m_2+1)th derivative of which is the function $x \mapsto \frac{1}{p!}\, x^{m_1+p}\,(1-x)^p$, characterized by the condition that its derivatives of a lesser order vanish at 0: hence,

$$S_p(m_1,m_2) = \frac{1}{p!\,m_2!} \int_0^1 x^{m_1+p}\,(1-x)^{m_2+p}\,dx$$

$$= \frac{(m_1+p)!\,(m_2+p)!}{p!\,m_2!\,(m_1+m_2+1+2p)!}, \qquad (3.4.56)$$

which leads to $\Gamma_{m_1+m_2+2+2p}^{m_1+1,m_2+1} = (\frac{i}{\pi})^p$.

Finally, that h lies in $\mathcal{G}_{m_1+m_2+1}(\mathbb{R}^2)$ is a consequence of Lemma 3.4.3, since we can now write

$$h(x,\xi) = \pi\,(x-i\xi)^{m_1+m_2+1} \sum_{p\geq 1} \frac{1}{p!}$$

$$\int_0^1 (1-t)^{m_1+p}\, t^{m_2+p} \left(\frac{x-i\xi}{x+i\xi}\,\frac{d}{dt}\right)^p [k^1((1-t)\,(x^2+\xi^2))\, k^2(t\,(x^2+\xi^2))]\,dt,$$

$$(3.4.57)$$

with k^1 and k^2 as defined in (3.4.25). One can then use Cauchy's formula for the estimation of derivatives of functions in a space such as $\mathcal{G}_{m_1}(\mathbb{R}^2)$ or do the following, which has the advantage of making the summation explicit: with $x+i\xi = r\,e^{i\phi}$, one may set $t = \frac{s}{1+s}$ in the integral on the right-hand side of (3.4.57) and, in the result, perform the change of complex contour $s \mapsto s\,\exp(e^{2i\phi})$, which leads to the formula

$$h(r\cos\phi, r\sin\phi) = \pi\,(re^{-i\phi})^{m_1+m_2+1}\,\exp\left(e^{2i\phi}\right) \times \int_0^\infty$$

$$[1+s\exp\left(e^{2i\phi}\right)]^{-m_1-2}\,[1+s^{-1}\exp\left(-e^{2i\phi}\right)]^{-m_2}\,k^1\left(\frac{x^2+\xi^2}{1+s}\right)\,k^2\left(\frac{x^2+\xi^2}{1+s^{-1}}\right)\,ds.$$

$$(3.4.58)$$

$$\square$$

Remark 3.4.1. Under the wording of *star-product* theory, an axiomatization [1], then a general construction [17] somewhat related to the sharp products of pseudodifferential analysis, have been put forward: we take this opportunity to emphasize to which extent our point of view, identifying quantization with pseudodifferential analysis, lies away from this concept. Star products are what remains, or a generalization of what remains, from a pseudodifferential analysis when the *right-hand side* only of the composition formula is remembered: actually, in the deformation point of view, only the formal expansion in terms of a series of bidifferential operators is taken into account. The sole demand is that the star product should be associative, not that it should correspond to the composition of operators in some symbolic calculus.

In Sect. 4.2, it will be shown that the sharp composition formula can remain the same for a one-parameter family of inequivalent pseudodifferential analyses: hence, it cannot, even theoretically, provide as much information as a genuine quantization theory (i.e., pseudodifferential analysis). On the other hand, in [36], it has been shown that, in the usual Weyl calculus, Moyal-type expansions are very far from revealing the general aspects of the sharp composition formula. Indeed, there exist pairs of (Eisenstein) distributions $\mathfrak{E}^\#_{i\lambda_1}$ and $\mathfrak{E}^\#_{i\lambda_2}$ giving rise to a fully explicit sharp composition, even though not one term from the would-be Moyal expansion, for instance the pointwise product, could make sense. This shows that the deformation point of view breaks down completely in some important instances.

Our point of view in covariant quantization theory, developed to some extent in [36, Sect. 19], is that, not even touching upon the immense applications of pseudodifferential analysis to partial differential equations and to mathematical physics, one should conceive of quantization theory as a meeting ground of representation theory and spectral theory. Then, series (convergent or not) the deformation point of view leads to only appear as very special cases of the expansions one gets when coupling the sharp composition of symbols with their decomposition relative to the action on symbols of operators in the covariance group of the calculus. This scheme (which we refrained from axiomatizing: quantization theory seems to offer consistently new models which escape previous concepts) makes it possible to consider integrals in place of series, and when needed, which is the case in automorphic pseudodifferential analysis, combinations of series and integrals.

Chapter 4
From Anaplectic Analysis to Usual Analysis

It is possible to consider anaplectic analysis on the real line as a special case of a one-parameter family of analyses. The parameter v is a complex number mod 2, subject to the restriction that it should not be an integer: anaplectic analysis, as considered until now, corresponds to the case when $v = -\frac{1}{2}$. There is a natural v-anaplectic representation of some cover of $SL(2, \mathbb{R})$ in some space $\mathfrak{A}_v$, compatible in the usual way with the Heisenberg representation; the v-anaplectic representation is pseudo-unitarizable in the case when v is real. Depending on v, the even or odd part of the v-anaplectic representation coincides in this case with a representation taken from the unitary dual of the universal cover of $SL(2, \mathbb{R})$, as completely described by Pukanszky [24]. In v-anaplectic analysis, the spectrum of the harmonic oscillator is the arithmetic sequence $v + \frac{1}{2} + \mathbb{Z}$. Much of the theory subsists in the case when $v \equiv 0 \mod 2$, which leads to a nontrivial enlargement of usual analysis. However, as we shall make clear, while the ascending pseudodifferential calculus extends to the case of v-anaplectic analysis when $v \in \mathbb{C} \backslash \mathbb{Z}$, it is impossible to extend it to the usual analysis environment.

4.1 The v-Anaplectic Representation

The one-parameter generalization of anaplectic analysis to be summed up in the present section was introduced in [38, Sects. 11–12]. The most natural way to characterize the space $\mathfrak{A}_v$ of functions on the real line which is basic in v-anaplectic analysis is probably that which follows from a generalization of Proposition 2.2.11: this characterization will be given in Theorem 4.1.3. However, we prefer to follow the plan of Sect. 2.2, starting from the $\mathbb{C}^4$-realization of functions in $\mathfrak{A}_v$.

Definition 4.1.1. Let $v \in \mathbb{C} \backslash \mathbb{Z}$ and consider the space of $\mathbb{C}^4$-valued functions $f = (f_0, f_1, f_{i,0}, f_{i,1})$ with the following properties: each component of f is a nice function in the sense of Definition 2.2.1, and the components are linked by the following equations:

$$f_{i,0}(x) = \frac{\Gamma(-v)}{(2\pi)^{\frac{1}{2}}} \left[e^{-\frac{i\pi}{2}(v+1)} f_0(ix) + e^{\frac{i\pi}{2}(v+1)} f_0(-ix) \right],$$

$$f_{i,1}(x) = \frac{\Gamma(-v)}{(2\pi)^{\frac{1}{2}}} \left[e^{-\frac{i\pi v}{2}} f_1(ix) + e^{\frac{i\pi v}{2}} f_1(-ix) \right]. \tag{4.1.1}$$

The space $\mathfrak{A}_v$ is the image of the space of functions so defined under the map $f \mapsto u$, where the even (resp., odd) part of u is the even part of f_0 (resp., the odd part of f_1). We shall also refer to f as the $\mathbb{C}^4$-realization of u.

Remark 4.1.1. The space $\mathfrak{A}_v$ only depends on v mod 2: for if $(f_0, f_1, f_{i,0}, f_{i,1})$ is a $\mathbb{C}^4$-realization of u when the parameter v is considered, the vector $h = (f_0, f_1, h_{i,0}, h_{i,1})$, with

$$h_{i,0} = -((v+1)(v+2))^{-1} f_{i,0}, \qquad h_{i,1} = -((v+1)(v+2))^{-1} f_{i,1}, \tag{4.1.2}$$

is a $\mathbb{C}^4$-realization of u in the space $\mathfrak{A}_{v+2}$. When using $\mathbb{C}^4$-realizations, we shall always assume that a value of $v \in \mathbb{C} \backslash \mathbb{Z}$ has been fixed.

Again, the Phragmén–Lindelöf lemma makes it possible to show that the map $f \mapsto u$ is one to one. Unless $v \in -\frac{1}{2} + \mathbb{Z}$, the space $\mathfrak{A}_v$ is not invariant under the complex rotation by 90°. However, the following holds.

Proposition 4.1.2. *The map $u \mapsto u_i$, with $u_i(x) = u(ix)$, is a linear isomorphism from $\mathfrak{A}_v$ to $\mathfrak{A}_{-v-1}$. If $(f_0, f_1, f_{i,0}, f_{i,1})$ is the $\mathbb{C}^4$-realization of u in the $\mathfrak{A}_v$-analysis, that of u_i in the $\mathfrak{A}_{-v-1}$-analysis is*

$$(h_0, h_1, h_{i,0}, h_{i,1}) = C_v (f_{i,0}, -i f_{i,1}, f_0, -i f_1,) \tag{4.1.3}$$

with

$$C_v = 2^{v+\frac{1}{2}} \frac{\Gamma(\frac{2+v}{2})}{\Gamma(\frac{1-v}{2})}. \tag{4.1.4}$$

The Heisenberg transformations $e^{2i\pi(\eta Q - yP)}$, with $(y, \eta) \in \mathbb{C}^2$, preserve the space $\mathfrak{A}_v$.

Theorem 4.1.3. *Let u be an entire function of one variable satisfying for some pair of constants C, R the estimate $|f(z)| \leq C e^{\pi R |z|^2}$. Define the functions $(Qu)_j$ and $(Ku)_j$ ($j = 0$ or 1) in the same way as in Proposition 2.2.11. Given $v \in \mathbb{C} \backslash \mathbb{Z}$, the following three conditions are equivalent:*

(i) u lies in the space $\mathfrak{A}_v$;

(ii) each of the two functions $(Qu)_0$ and $(Qu)_1$ extends as an analytic function on the real line, admitting for large $|\sigma|$ a convergent expansion $(Qu)_j(\sigma) = e^{-\frac{i\pi}{2}(v+\frac{1}{2})\operatorname{sign}\sigma} \sum_{n \geq 0} a_n^{(j)} \sigma^{-n} |\sigma|^{-\frac{1}{2}}$;

(iii) each of the two functions $(Ku)_0$ and $(Ku)_1$, initially defined in a neighborhood of the point $z = 1$ of the unit circle S^1, extends as an analytic function to the universal cover of S^1, satisfying the quasiperiodicity conditions

$$(\mathcal{K}u)_j(e^{i\theta}) = e^{-i\pi(\nu+\frac{1}{2})}(\mathcal{K}u)_j(e^{i(\theta-2\pi)}). \qquad (4.1.5)$$

We now consider the ν-analogue of Theorem 2.2.3, searching for the eigenfunctions of the standard harmonic oscillator $L = \pi(Q^2 + P^2)$ which lie in $\mathfrak{A}_\nu$. As a consequence of the WKB method, near each of the two endpoints $\pm\infty$ of the real line, the equation $Lf = (\nu + \frac{1}{2})f$ has two solutions: one that behaves like $|x|^\nu e^{-\pi x^2}$ and another like $|x|^{-\nu-1} e^{\pi x^2}$. Since we are assuming that $\nu \neq 0, 1, \dots$, no solution can be rapidly decreasing toward $\pm\infty$ simultaneously. In [21, Chap. 8], one denotes as $f(x) = D_\nu(2\pi^{\frac{1}{2}}x)$ the solution of the equation $Lf = (\nu + \frac{1}{2})f$ normalized by the condition

$$f(x) \sim (2\pi^{\frac{1}{2}}x)^\nu e^{-\pi x^2}, \qquad x \to +\infty: \qquad (4.1.6)$$

of course, such a function is very far from lying in $L^2(\mathbb{R})$.

The following definition generalizes Proposition 2.2.2: in the case when $\nu = -\frac{1}{2}$, the function $\psi^{-\frac{1}{2}}$ to be introduced now and the function ϕ introduced there only differ by some normalizing factor. That the $\mathbb{C}^4$-valued function below is indeed the $\mathbb{C}^4$-realization of some function in $\mathfrak{A}_\nu$, i.e., that the equations (4.1.1) are valid, is a consequence of [21, p. 330].

Proposition 4.1.4. *Let $\nu \notin \mathbb{Z}$ and let ψ^ν be the function in $\mathfrak{A}_\nu$ the $\mathbb{C}^4$-realization of which is the function*

$$f(x) = (D_\nu(2\pi^{\frac{1}{2}}x), 0, D_{-\nu-1}(2\pi^{\frac{1}{2}}x), 0). \qquad (4.1.7)$$

One has, for $0 < \theta < 2\pi$,

$$(\mathcal{K}\psi^\nu)_0(e^{-i\theta}) = \frac{2^{\frac{\nu-1}{2}}\pi^{\frac{1}{2}}}{\Gamma(\frac{1-\nu}{2})} e^{\frac{i}{2}(\nu+\frac{1}{2})\theta}. \qquad (4.1.8)$$

Theorem 4.1.5. *The set of eigenvalues of the harmonic oscillator in the space $\mathfrak{A}_\nu$ is the arithmetic sequence $\nu + \frac{1}{2} + \mathbb{Z}$. The eigenspace corresponding to an eigenvalue $\nu + 2j + \frac{1}{2}$, $j \in \mathbb{Z}$, is generated by the function $\psi^{\nu+2j}$ with the $\mathbb{C}^4$-realization*

$$x \mapsto \left(D_{\nu+2j}(2\pi^{\frac{1}{2}}x), 0, (-1)^j \frac{\Gamma(\nu+2j+1)}{\Gamma(\nu+1)} D_{-\nu-2j-1}(2\pi^{\frac{1}{2}}x), 0\right) \qquad (4.1.9)$$

and the eigenspace corresponding to an eigenvalue $\nu + 2j + \frac{3}{2}$, $j \in \mathbb{Z}$, is generated by the function $\chi^{\nu+2j+1}$ with the $\mathbb{C}^4$-realization

$$x \mapsto \left(0, D_{\nu+2j+1}(2\pi^{\frac{1}{2}}x), 0, (-1)^j \frac{\Gamma(\nu+2j+2)}{\Gamma(\nu+1)} D_{-\nu-2j-2}(2\pi^{\frac{1}{2}}x)\right). \qquad (4.1.10)$$

One has the relations

$$A\psi^{\nu+2j} = (\nu+2j)\chi^{\nu+2j-1}, \qquad A^*\psi^{\nu+2j} = \chi^{\nu+2j+1},$$
$$A\chi^{\nu+2j+1} = (\nu+2j+1)\psi^{\nu+2j}, \qquad A^*\chi^{\nu+2j+1} = \psi^{\nu+2j+2}. \qquad (4.1.11)$$

Remark 4.1.2. By its definition in Proposition 4.1.4, the function $\psi^{\nu+2j}$ lies in $\mathfrak{A}_{\nu+2j} = \mathfrak{A}_\nu$: however, as seen from Remark 4.1.1, its $\mathbb{C}^4$-realization is not the same in the ν-anaplectic or in the $(\nu+2j)$-anaplectic theory. One should not make a confusion between the odd function $\chi^{\nu+2j+1} \in \mathfrak{A}_\nu$ and the even function $\psi^{\nu+2j+1}$, which does not lie in $\mathfrak{A}_\nu$ but in $\mathfrak{A}_{\nu+1}$, and which will not concern us; in analogy with (4.1.8), one has the equation

$$(\mathcal{K}\chi^{\nu+1})_1(e^{-i\theta}) = e^{\frac{i\pi}{4}} \frac{2^{\frac{\nu-1}{2}}(\nu+1)}{\Gamma(\frac{1-\nu}{2})} e^{\frac{i}{2}(\nu+\frac{1}{2})\theta}. \tag{4.1.12}$$

Equation (11.56) given in [38, Theorem 11.10] is erroneous and should be replaced by the present equation (4.1.12). It is the equation

$$\pi^{\frac{1}{2}} x^2 - \frac{1}{2\pi^{\frac{1}{2}}} x \frac{d}{dx} = \frac{\pi^{-\frac{1}{2}}}{2} [L + A^{*2} + \frac{1}{2}] \tag{4.1.13}$$

which must take the place of equation (11.58) given there.

Remark 4.1.3. Since [21, p. 326] $D_{-\frac{1}{2}}(2\pi^{\frac{1}{2}}x) = \pi^{-\frac{1}{4}} x^{\frac{1}{2}} K_{\frac{1}{4}}(\pi x^2)$ for $x > 0$, one sees, comparing $\psi^{-\frac{1}{2}}$ with the function ϕ introduced in Proposition 2.2.2 and using the analytic continuation of the function $K_{\frac{1}{4}}$ as provided, for instance, by [21, p. 69], that $\phi = 2^{\frac{1}{2}} \pi^{-\frac{1}{4}} \psi^{-\frac{1}{2}}$. More generally, the function $\psi^{-\frac{1}{2}+2j}$ is a multiple of the function denoted as ϕ^{2j} in (2.2.10) and the function $\chi^{-\frac{1}{2}+2j+1}$ is a multiple of the function ϕ^{2j+1}. The coefficients of proportionality can be obtained from (4.1.11) and Lemma 2.2.9.

Contrary to that of the $\mathbb{C}^4$-realization of a function in $\mathfrak{A}_\nu$, Definition 4.1.6 depends only on ν mod 2.

Definition 4.1.6. Given $u \in \mathfrak{A}_\nu$, we set

$$\mathrm{Int}[u] = e^{\frac{i\pi}{2}(\nu+\frac{1}{2})} \left[2\cos\frac{\pi\nu}{2} \int_0^\infty f_0(x)\,dx + \frac{(2\pi)^{\frac{1}{2}}}{\Gamma(-\nu)} \int_0^\infty f_{i,0}(x)\,dx \right]. \tag{4.1.14}$$

The ν-anaplectic Fourier transformation $\mathcal{F}_{\mathrm{ana}}^\nu$ is defined by the equation

$$(\mathcal{F}_{\mathrm{ana}}^\nu u)(x) = \mathrm{Int}[y \mapsto e^{-2i\pi xy} u(y)], \qquad u \in \mathfrak{A}_\nu. \tag{4.1.15}$$

Theorem 4.1.7. *The linear form* Int *is invariant under the (real or complex) Heisenberg translations* $\pi(y,0)$, *and the ν-anaplectic Fourier transformation is a linear automorphism of the space* $\mathfrak{A}_\nu$. *In terms of the $\mathcal{K}$-transform of u, one has*

$$\mathrm{Int}[u] = 2^{\frac{1}{2}} (\mathcal{K}u)_0(e^{-i\pi}) \tag{4.1.16}$$

and the equations

$$(\mathcal{K}(\mathcal{F}_{\mathrm{ana}}^\nu u))_0(z) = e^{i\pi(\nu+\frac{1}{2})} (\mathcal{K}u)_0(e^{i\pi} z),$$
$$(\mathcal{K}(\mathcal{F}_{\mathrm{ana}}^\nu u))_1(z) = e^{i\pi\nu} (\mathcal{K}u)_1(e^{i\pi} z). \tag{4.1.17}$$

For every $v \in \mathbb{C}\backslash\mathbb{Z}$, there is a v-version of the anaplectic representation. The following generalizes Theorem 2.2.3: however, unless $v \in -\frac{1}{2} + \mathbb{Z}$, it is necessary to substitute for $SL(2,\mathbb{R})$ a cover of this group to get a genuine representation. Note that elements such as $\begin{pmatrix} 1 & 0 \\ c & 1 \end{pmatrix}$ or $\begin{pmatrix} a & 0 \\ 0 & a^{-1} \end{pmatrix}$ of $SL(2,\mathbb{R})$ are naturally associated to elements, denoted in the same way, of the universal cover of $SL(2,\mathbb{R})$; in what follows, $\exp \frac{\pi}{2} \begin{pmatrix} 0 & 1 \\ -1 & 0 \end{pmatrix}$ denotes the element of the cover $G^{(N)}$ under consideration, which is the end of the path originating at the identity and covering the path $t \mapsto \begin{pmatrix} \cos t & \sin t \\ -\sin t & \cos t \end{pmatrix}$ of $SL(2,\mathbb{R})$.

Theorem 4.1.8. *Assume that $N = \infty$ or that N is a positive integer such that $N \left(\frac{v}{2} + \frac{1}{4} \right) \in \mathbb{Z}$. There exists a unique representation Ana_v of the N-fold cover $G^{(N)}$ of $G = SL(2,\mathbb{R})$ in the space $\mathfrak{A}_v$ with the following properties:*

(i) if $g = \begin{pmatrix} 1 & 0 \\ c & 1 \end{pmatrix}$, one has $(\mathrm{Ana}_v(g)u)(x) = u(x) e^{i\pi cx^2}$;

(ii) if $g = \begin{pmatrix} a & 0 \\ 0 & a^{-1} \end{pmatrix}$ with $a > 0$, one has $(\mathrm{Ana}_v(g)u)(x) = a^{-\frac{1}{2}} u(a^{-1}x)$;

(iii) one has $\mathrm{Ana}_v \left(\exp \frac{\pi}{2} \begin{pmatrix} 0 & 1 \\ -1 & 0 \end{pmatrix} \right) = e^{-i\pi(v+\frac{1}{2})} \mathcal{F}_{\mathrm{ana}}^v$.

This representation combines with the Heisenberg representation in the way expressed by (2.2.18), only replacing Ana by Ana_v.

The v-anaplectic representation can be defined globally with the help of $\mathcal{K}$-transforms of functions in Ana_v. Assume that $g \in G^{(N)}$ lies above the matrix $\begin{pmatrix} a & b \\ c & d \end{pmatrix} \in G$, and set $\begin{pmatrix} \alpha & \beta \\ \bar{\beta} & \bar{\alpha} \end{pmatrix} = S \begin{pmatrix} a & b \\ c & d \end{pmatrix} S^{-1}$ with $S = 2^{-\frac{1}{2}} \begin{pmatrix} 1 & i \\ i & 1 \end{pmatrix}$. On the other hand, associate with $\begin{pmatrix} a & b \\ c & d \end{pmatrix}$ the matrix

$$\begin{pmatrix} \lambda & \mu \\ \bar{\mu} & \bar{\lambda} \end{pmatrix} = \frac{1}{2} \begin{pmatrix} a - ib + ic + d & -a - ib - ic + d \\ -a + ib + ic + d & a + ib - ic + d \end{pmatrix} \tag{4.1.18}$$

and the transformation $z \mapsto \frac{\lambda z + \mu}{\bar{\mu} z + \bar{\lambda}}$ of S^1: this extends as a homomorphism $g \mapsto [g]$ of $G^{(N)}$ into the group of analytic automorphisms of the N-fold cover $\Sigma^{(N)}$ of $\Sigma = S^1$. Denoting as $\frac{[g^{-1}]_* d\theta}{d\theta}$ the Radon–Nikodym derivative of the transformation $[g^{-1}]$ with respect to the "rotation"-invariant measure of $\Sigma^{(N)}$, one can characterize the v-anaplectic representation by the pair of equations

$$(\mathcal{K} \mathrm{Ana}_v(g)u)_0(z) = \left(\frac{[g^{-1}]_* d\theta}{d\theta}(z) \right)^{\frac{1}{4}} (\mathcal{K}u)_0([g^{-1}](z)),$$

$$(\mathcal{K} \mathrm{Ana}_v(g)u)_1(z) = \left(\frac{[g^{-1}]_* d\theta}{d\theta}(z) \right)^{\frac{1}{4}} \left[\alpha - i\beta \left([g^{-1}](z)\right)^{-1} \right] . (\mathcal{K}u)_1([g^{-1}](z)).$$

$$\tag{4.1.19}$$

Finally, when v is real (and not an integer), there is on $\mathfrak{A}_v$ a nondegenerate pseudoscalar product invariant under the v-anaplectic representation as well as under the Heisenberg representation (considering this time only the operators $e^{2i\pi(\eta Q - yP)}$ with $(y, \eta) \in \mathbb{R}^2$).

Theorem 4.1.9. *Let* $v \in \mathbb{R} \backslash \mathbb{Z}$. *The* v*-anaplectic representation and the Heisenberg representation are both pseudo-unitary with respect to the pseudoscalar product on* Ana_v *defined, in terms of the* $\mathbb{C}^4$*-realization* $(f_0, f_1, f_{i,0}, f_{i,1})$ *of* u, *by the equation*

$$(u|u) = 2^{\frac{1}{2}} \int_0^\infty \left[|f_0|^2 + |f_1|^2 + \frac{\Gamma(v+1)}{\Gamma(-v)} (|f_{i,0}|^2 - |f_{i,1}|^2) \right] dx. \qquad (4.1.20)$$

This pseudoscalar product is nondegenerate and only depends on v *mod* 2. *In terms of the* $\mathcal{K}$*-realization, setting*

$$(\mathcal{K}u)_0(z) = z^{-\frac{1}{2}(v+\frac{1}{2})} \sum_{j \in \mathbb{Z}} c_j z^{-j},$$

$$(\mathcal{K}u)_1(z) = z^{-\frac{1}{2}(v+\frac{1}{2})} \sum_{j \in \mathbb{Z}} c'_j z^{-j}, \qquad (4.1.21)$$

one has

$$(u|u) = \frac{\pi^{\frac{1}{2}}}{\cos^2 \frac{\pi v}{2}} \sum_{j \in \mathbb{Z}} \Gamma(\frac{v}{2} + j + 1) \left[\frac{2}{\Gamma(\frac{v+1}{2} + j)} |c_j|^2 + \frac{\pi}{2\Gamma(\frac{v+3}{2} + j)} |c'_j|^2 \right]. \qquad (4.1.22)$$

In particular,

$$(\psi^{v+2j} | \psi^{v+2j}) = \frac{1}{2} \Gamma(v + 2j + 1),$$

$$(\chi^{v+2j+1} | \chi^{v+2j+1}) = \frac{1}{2} \Gamma(v + 2j + 2). \qquad (4.1.23)$$

Proof. There would be no need to redo the proof, in principle given in [38] as the proof of Theorem 12.4 there. However, the second term of the sum on the right-hand side of (4.1.22) does not agree with the one given there. This is the only place where we had to use the former version of (4.1.12) which, as pointed out in Remark 4.1.2, was erroneous. Now, the coefficients of the identity under study have been defined, in loc.cit., so as to make (4.1.23) valid: this leads to the corrected version (4.1.22). □

Remark 4.1.4. (i) That the pseudoscalar product only depends on v mod 2 may seem surprising: only, do not forget (cf. (4.1.2)) that the $\mathbb{C}^4$-realization of a function $u \in \mathfrak{A}_v$ depends on v, not only on v mod 2.

(ii) In the case when $v \in]-1, 0[+2\mathbb{Z}$, the pseudoscalar product is positive definite when restricted to the even part of $\mathfrak{A}_v$; when $v \in]0, 1[+2\mathbb{Z}$, it is positive definite on the odd part of this space.

(iii) In Proposition 2.2.12, it has been shown that the even part of the anaplectic representation is unitarily equivalent to some representation of $SL(2, \mathbb{R})$ taken from the complementary series of that group. It can be shown that, in the case when $v \in]-1, 0[+2\mathbb{Z}$ (resp. $v \in]0, 1[+2\mathbb{Z}$), the restriction of the v-anaplectic representation to the even (resp., odd) part of Ana_ν coincides with a representation taken from the unitary dual of the universal cover of $SL(2,\mathbb{R})$ [24]: details can be found in the last section of [38].

(iv) As seen from (4.1.8) and (4.1.17), one has

$$\mathcal{F}_{\text{ana}}^\nu \, \psi^\nu = e^{\frac{i\pi}{2}(\nu+\frac{1}{2})} \, \psi^\nu :$$ (4.1.24)

in view of the condition (iii) from Theorem 4.1.8 and of (2.2.30), still valid when Ana_ν is substituted for Ana provided that one interprets the matrix $\left(\begin{smallmatrix} \cos t & \sin t \\ -\sin t & \cos t \end{smallmatrix}\right)$ as $\exp t \left(\begin{smallmatrix} 0 & 1 \\ -1 & 0 \end{smallmatrix}\right)$, one obtains

$$\exp\left(-\frac{i\pi}{2} L\right) \psi^\nu = e^{-\frac{i\pi}{2}(\nu+\frac{1}{2})} \, \psi^\nu,$$ (4.1.25)

a form of the eigenvalue equation $L \psi^\nu = (\nu + \frac{1}{2}) \psi^\nu$.

More generally, for any point z in the upper half-plane Π, we consider (cf. Proposition 2.2.8) the operators $A_z = \pi^{\frac{1}{2}} (Q - \bar{z}P)$ and $L_z = A_z A_z^* - \frac{\text{Im} z}{2}$. Let $\left(\begin{smallmatrix} a & b \\ c & d \end{smallmatrix}\right) \in SL(2,\mathbb{R})$ be given and let g be any point of the group $G^{(N)}$ introduced in Theorem 4.1.8 lying above this matrix: one then has

$$\text{Ana}_\nu(g) A_z \text{Ana}_\nu(g^{-1}) = (c\bar{z}+d) A_{\frac{az+b}{cz+d}},$$
$$\text{Ana}_\nu(g) L_z \text{Ana}_\nu(g^{-1}) = |c\bar{z}+d|^2 L_{\frac{az+b}{cz+d}}.$$ (4.1.26)

These identities, the ν-anaplectic analogue of (2.2.22), can be proved in the same way, as a consequence of the analogue of (2.2.19).

If $z = x + iy \in \Pi$ and g_z is a point of $G^{(N)}$ above the matrix $\left(\begin{smallmatrix} y^{\frac{1}{2}} & y^{-\frac{1}{2}}x \\ 0 & y^{-\frac{1}{2}} \end{smallmatrix}\right)$, finally if one sets

$$\phi_z^{\nu,k} = \begin{cases} \text{Ana}_\nu(g_z) \, \psi^{\nu+k} & \text{if } k \text{ is even} \\ \text{Ana}_\nu(g_z) \, \chi^{\nu+k} & \text{if } k \text{ is odd,} \end{cases}$$ (4.1.27)

one obtains a full set of ν-anaplectic eigenfunctions of L, $\phi_z^{\nu,k}$ corresponding to the eigenvalue $\nu + k$. However, defining g_z without any ambiguity requires more care. Set

$$z = \begin{pmatrix} \alpha & 0 \\ \gamma & \alpha^{-1} \end{pmatrix} . i = \begin{pmatrix} \alpha & 0 \\ \gamma & \alpha^{-1} \end{pmatrix} \begin{pmatrix} \frac{y}{|z|} & \frac{x}{|z|} \\ -\frac{x}{|z|} & \frac{y}{|z|} \end{pmatrix} . i :$$ (4.1.28)

since the last matrix on the right-hand side is orthogonal, its associated fractional-linear transformation fixes the point i, so that this equation is valid provided that $z = \frac{\alpha^2}{\alpha\gamma-i}$, in other words $\alpha = y^{-\frac{1}{2}}|z|$, $\gamma = y^{-\frac{1}{2}}|z|^{-1}x$.

To define g_z, it suffices to cover each of the two matrices on the right-hand side of (4.1.28) by a well-defined element of the group $G^{(N)}$. For the first one, it is obvious how to do this, only splitting it as the product $\left(\begin{smallmatrix} 1 & 0 \\ \gamma\alpha^{-1} & 1 \end{smallmatrix}\right) \left(\begin{smallmatrix} \alpha & 0 \\ 0 & \alpha^{-1} \end{smallmatrix}\right)$: the associated element of the ν-anaplectic representation is the map $u \mapsto v$ with

$$v(t) = \alpha^{-\frac{1}{2}} e^{i\pi\frac{\gamma}{\alpha}t^2} u(\alpha^{-1}t).$$ (4.1.29)

Since $y > 0$, one can uniquely set $y + ix = e^{i\theta}$ with $|\theta| < \frac{\pi}{2}$, which gives a meaning to fractional powers of $\frac{-iz}{|z|} = e^{-i\theta}$. Writing the last matrix on the right-hand side of (4.1.28) as $\exp \theta \left(\begin{smallmatrix} 0 & 1 \\ -1 & 0 \end{smallmatrix} \right)$, one can regard it as an element of $G^{(N)}$. To see the effect of the associated element of the ν-anaplectic representation on $\psi^{\nu+k}$ (or $\chi^{\nu+k}$, depending on the parity of k), we use the equation

$$\exp(-i\theta L) = \mathrm{Ana}_\nu \left(\exp \theta \left(\begin{smallmatrix} 0 & 1 \\ -1 & 0 \end{smallmatrix} \right) \right), \tag{4.1.30}$$

the proof of which is absolutely identical to that of (2.2.30). Since (for even k) $L\psi^{\nu+k} = (\nu + \frac{1}{2} + k)\,\psi^{\nu+k}$, we finally obtain

$$\phi_z^{\nu,k}(t) = \left(\frac{-iz}{|z|} \right)^{\nu+\frac{1}{2}+k} \frac{(\mathrm{Im}\, z)^{\frac{1}{4}}}{|z|^{\frac{1}{2}}}\, \psi^{\nu+k} \left(\frac{(\mathrm{Im}\, z)^{\frac{1}{2}} t}{|z|} \right) e^{i\pi \frac{\mathrm{Re}\, z}{|z|^2} t^2} \tag{4.1.31}$$

if k is even, and a fully similar equation, only replacing $\psi^{\nu+k}$ by $\chi^{\nu+k}$, if k is odd.

We finally generalize Lemma 2.2.9.

Lemma 4.1.10. *For every $z \in \Pi$ and $k \in \mathbb{Z}$, one has*

$$A_z\,\phi_z^{\nu,k} = (\nu + k)\,(\mathrm{Im}\, z)^{\frac{1}{2}}\,\phi_z^{\nu,k-1}, \qquad A_z^*\,\phi_z^{\nu,k} = (\mathrm{Im}\, z)^{\frac{1}{2}}\,\phi_z^{\nu,k+1}. \tag{4.1.32}$$

Also,

$$\left[4i\,(\mathrm{Im}\, z)\frac{\partial}{\partial z} - \nu - \frac{1}{2} - k \right]\phi_z^{\nu,k} = -(\nu+k)(\nu+k-1)\,\phi_z^{\nu,k-2},$$

$$\left[4i\,(\mathrm{Im}\, z)\frac{\partial}{\partial \bar{z}} - \nu - \frac{1}{2} - k \right]\phi_z^{\nu,k} = -\phi_z^{\nu,k+2}. \tag{4.1.33}$$

Proof. The proof of the first part is based on (4.1.11) and (4.1.26). Note that, in the case when $\nu = -\frac{1}{2}$, comparing this pair of equations with the corresponding pair (2.2.34) from Lemma 2.2.9 makes it possible to obtain the coefficients of proportionality between the functions $\phi_z^{-\frac{1}{2},k}$ and ϕ_z^k, starting from Remark 4.1.3.

For the second part, we still follow the proof of Lemma 2.2.9, interpreting this time J as the element $\exp\frac{\pi}{2}\left(\begin{smallmatrix} 0 & 1 \\ -1 & 0 \end{smallmatrix} \right)$ of $G^{(N)}$ rather than as a matrix. Assuming for instance that k is even (if not, it suffices to change ψ to χ), (2.2.41) becomes

$$A^2\,\psi^{\nu+k} = 2\pi Q\,(Q+iP)\,\psi^{\nu+k} - (\nu+k)\,\psi^{\nu+k},$$

$$A^{*2}\,\psi^{\nu+k} = 2\pi Q\,(Q-iP)\,\psi^{\nu+k} - (\nu+1+k)\,\psi^{\nu+k}, \tag{4.1.34}$$

and (2.2.43) becomes

$$\mathrm{Ana}_\nu(\tilde{g}_z)\,A^2\,\psi^{\nu+k} = \left(4iy\frac{\partial}{\partial z} - \nu - \frac{1}{2} - k \right)\mathrm{Ana}_\nu(\tilde{g}_z)\,\psi^{\nu+k},$$

$$\mathrm{Ana}_\nu(\tilde{g}_z)\,A^{*2}\,\psi^{\nu+k} = \left(4iy\frac{\partial}{\partial \bar{z}} - \nu - \frac{1}{2} - k \right)\mathrm{Ana}_\nu(\tilde{g}_z)\,\psi^{\nu+k}. \tag{4.1.35}$$

Also, one has

$$\mathrm{Ana}_v(J)\,\psi^{v+k} = e^{-\frac{i\pi}{2}(v+\frac{1}{2}+k)}\,\psi^{v+k}. \tag{4.1.36}$$

Equations (4.1.33) follow. $\qquad\qquad\qquad\qquad\qquad\qquad\qquad\qquad\qquad\qquad\square$

4.2 Ascending Pseudodifferential Calculus in v-Anaplectic Analysis

We here briefly show how the ascending pseudodifferential analysis extends to the v-anaplectic case, under the assumption that $v \in \mathbb{R}\backslash\mathbb{Z}$. Then, we shall examine which part of v-anaplectic analysis can survive when $v \equiv 0 \bmod 2$, and why this case contains usual analysis.

As a first step, we start with a generalization of Proposition 3.3.1, to the effect that the operator $Q+iP$ is an automorphism of $\mathfrak{A}_v$ for every $v \in \mathbb{C}\backslash\mathbb{Z}$. This is proved with the help of the following formulas, in which $\boldsymbol{f} = (f_0, f_1, f_{i,0}, f_{i,1})$ is the $\mathbb{C}^4$-realization of some function $u \in \mathfrak{A}_v$ and $\boldsymbol{g} = (g_0, g_1, g_{i,0}, g_{i,1})$ is the $\mathbb{C}^4$-realization of the function $u_1 = (Q+iP)^{-1}u$:

$$g_0(x) = 2\pi \left[\int_0^x e^{-\pi(x^2-y^2)} f_1(y)\,dy - 2^{v+\frac{1}{2}} \frac{\Gamma(\frac{2+v}{2})}{\Gamma(\frac{1-v}{2})} e^{-\pi x^2} \int_0^\infty e^{-\pi y^2} f_{i,1}(y)\,dy \right],$$

$$g_{i,0}(x) = -2\pi e^{\pi x^2} \int_x^\infty e^{-\pi y^2} f_{i,1}(y)\,dy,$$

$$g_1(x) = 2\pi \left[\int_0^x e^{-\pi(x^2-y^2)} f_0(y)\,dy + 2^{v+\frac{1}{2}} \frac{\Gamma(\frac{1+v}{2})}{\Gamma(-\frac{v}{2})} e^{-\pi x^2} \int_0^\infty e^{-\pi y^2} f_{i,0}(y)\,dy \right],$$

$$g_{i,1}(x) = 2\pi e^{\pi x^2} \int_x^\infty e^{-\pi y^2} f_{i,0}(y)\,dy. \tag{4.2.1}$$

No change has to be done in the proof of Proposition 3.3.1, except for the use of (4.1.1) in place of (2.2.2).

We then extend Definition 3.1.2 to the environment of v-anaplectic analysis and start with observing that all of Sect. 3.1 extends without modification: indeed, the lemmas in this section are only concerned with analysis on Π, once the invertibility of A_z has been established.

Extending Sects. 3.2–3.4, however, requires a few inessential modifications. First, since pseudoscalar products play a basic role, we shall assume, so as to take advantage of (4.1.20), that v is real (and not an integer). That the operators Q and P are still self-adjoint comes from the fact that (2.2.6) and (2.2.7) extend without modification to v-anaplectic analysis. Lemma 3.2.1 extends as the equation

$$(A_z^{-m-1}\phi_\zeta^{v,k}\,|\,\phi_\zeta^{v,j}) = \overline{C_{v,m}^{j,k}}\,(\mathrm{Im}\,\zeta)^{\frac{m+1}{2}}\,(z-\zeta)^{\frac{-m-1+j-k}{2}}\,(z-\bar{\zeta})^{\frac{-m-1-j+k}{2}}, \tag{4.2.2}$$

and one has $C_{\nu,m}^{k+m+1,k} = \frac{1}{2}\Gamma(\nu+k+1)$. The proof of this latter equation is based on (4.1.32) and (4.1.20): note that $C_{-\frac{1}{2},m}^{j,k}$ is not the same as the coefficient $C_m^{j,k}$ from Sect. 3.2, in view of the different normalizations of the eigenfunctions of L.

It is not absolutely necessary to extend Lemmas 3.2.2 and 3.2.3, in which it would suffice, anyway, to change some coefficients. The essential Lemma 3.2.4, Proposition 3.2.5, Theorem 3.2.6, and Lemma 3.2.7 extend without any modification. Some is required, however, in the proof of Theorem 3.2.8, which depends on some explicit equalities involving the coefficients $F_m^{k+m+1,k}$ introduced in Lemma 3.2.2 and the coefficients γ_k, γ_k^* from Lemma 2.2.9. Setting $C_{\nu,m}^{j,k} = \frac{(-2i)^{m+1}}{m!} F_{\nu,m}^{j,k}$ so as to extend (3.2.12) and setting

$$\gamma_{\nu,k} = \nu+k, \qquad \gamma_{\nu,k}^* = 1 \tag{4.2.3}$$

so that the first part of Lemma 4.1.10 should extend the corresponding part of Lemma 2.2.9 with the same notation, we first have to verify, so as to generalize (3.2.75), that

$$\frac{F_{\nu,m_0}^{k+m_0+2,k+1}}{F_{\nu,m_0}^{k+m_0+1,k}} = \frac{\gamma_{\nu,k+1}}{\gamma_{\nu,k+m_0+1}^*}, \tag{4.2.4}$$

an immediate task. Following the proof of Theorem 3.2.8, we next come to (3.2.83), which must be generalized in the obvious way: since $\gamma_{\nu,j+1}\,\gamma_{\nu,j}^* = \nu+j+1$, this is again immediate. Equation (3.2.88) has to be changed in the obvious way, only inserting the subscript ν where needed: only, since (2.2.36) has to be replaced by (4.1.33), the right-hand side of the equation that replaces (3.2.88) is now (compare (3.2.90))

$$(\operatorname{Im}\zeta)^{\frac{1}{2}}\left([k+\nu+\frac{1}{2}+m_0+1+\alpha+\beta-4i\,(\operatorname{Im}\zeta)\frac{\partial}{\partial\bar{\zeta}}]\,\phi_\zeta^{\nu,k+m_0+1+\alpha+\beta}\,|\,C\phi_\zeta^{\nu,k}\right)$$
$$-(\operatorname{Im}\zeta)^{\frac{1}{2}}\left(\phi_\zeta^{k+\nu+1+m_0+1+\alpha+\beta}\,|\,C[k+\nu+\frac{1}{2}-4i\,(\operatorname{Im}\zeta)\frac{\partial}{\partial\bar{\zeta}}]\,\phi_\zeta^{\nu,k}\right). \tag{4.2.5}$$

Since the shift $k \mapsto k+\nu+\frac{1}{2}$ occurs on both sides, (3.2.91) does not have to be modified, so that Theorem 3.2.8 extends.

In view of the set of equations (4.2.1), some coefficients have to be modified in the formulas from Sect. 3.3. If $u_2 = A_z^{-2}u$, keeping the same notation as in Proposition 3.3.2, one sees that no modification whatsoever occurs in the equations for $g_{i,0}$ and $g_{i,1}$. However, the other two equations are to be replaced by the following:

$$g_0(x) = -\frac{4\pi}{\bar{z}^2}\left[\int_0^x (x-y)\,e^{\frac{i\pi}{\bar{z}}(x^2-y^2)}\,f_0(y)\,dy\right.$$
$$\left.+\int_0^\infty 2^{\nu+\frac{1}{2}}\left[\frac{\Gamma(\frac{1+\nu}{2})}{\Gamma(-\frac{\nu}{2})}x - \frac{\Gamma(\frac{2+\nu}{2})}{\Gamma(\frac{1-\nu}{2})}y\right]e^{\frac{i\pi}{\bar{z}}(x^2+y^2)}\,f_{i,0}(y)\,dy\right] \tag{4.2.6}$$

and

$$g_{\frac{1}{2}}(x) = -\frac{4\pi}{\bar{z}^2}\left[\int_0^x (x-y)\,e^{\frac{i\pi}{\bar{z}}(x^2-y^2)}f_1(y)\,dy\right.$$
$$\left. -\int_0^\infty 2^{\nu+\frac{1}{2}}\left[\frac{\Gamma(\frac{2+\nu}{2})}{\Gamma(\frac{1-\nu}{2})}x - \frac{\Gamma(\frac{1+\nu}{2})}{\Gamma(-\frac{\nu}{2})}y\right]e^{\frac{i\pi}{\bar{z}}(x^2+y^2)}f_{i,1}(y)\,dy\right]. \qquad (4.2.7)$$

Then,

$$u_2(x) = -\frac{4\pi}{\bar{z}^2}\left[\int_0^x (x-y)\,e^{\frac{i\pi}{\bar{z}}(x^2-y^2)}u(y)\,dy\right.$$
$$\left. -2^{\nu+\frac{1}{2}}\frac{\Gamma(\frac{2+\nu}{2})}{\Gamma(\frac{1-\nu}{2})}\int_0^\infty [y f_{i,0}(y)+x f_{i,1}(y)]\,e^{\frac{i\pi}{\bar{z}}(x^2+y^2)}\,dy\right]. \qquad (4.2.8)$$

Consequently, (3.3.21) becomes

$$(\mathrm{Op}_1^{\mathrm{asc}}(h)\,u)(x) = -\pi\left[\int_0^x (x-y)\,k(x^2-y^2)\,u(y)\,dy\right.$$
$$\left. -2^{\nu+\frac{1}{2}}\frac{\Gamma(\frac{2+\nu}{2})}{\Gamma(\frac{1-\nu}{2})}\int_0^\infty k(x^2+y^2)\,[y f_{i,0}(y)+x f_{i,1}(y)]\,dy\right], \qquad (4.2.9)$$

but there is nothing to change in the proof of Theorem 3.3.4.

Finally, the results of Sect. 3.4 concerning the composition formula remain valid in ν-anaplectic analysis. The coefficient $(\frac{i}{\pi})^p$ on the right-hand side of the main formula (3.4.38) does not depend on ν, since its value is based on the result of Lemma 3.4.4: now, in the proof of that lemma, the constant $C_m^{j,k}$, which is to be replaced by $C_{\nu,m}^{j,k}$, disappears since it occurs on both sides of the equation one is taking advantage of.

Remark 4.2.1. As recalled in Remark 4.1.4, the even (resp., odd) part of the ν-anaplectic representation (if $\nu \in\,]-1,0[\,+2\mathbb{Z}$, resp. $]0,1[\,+2\mathbb{Z}$) is unitarily equivalent to a representation taken from the complementary series of the universal cover of $SL(2,\mathbb{R})$. This is a two-parameter series [24], and we are here considering only a one-parameter subfamily. If one considers instead another one-parameter family, to wit that which corresponds to representations of $SL(2,\mathbb{R})$ (as opposed to the universal cover of that group), the situation is totally different. In that case, it is in general not possible to combine two representations (as we have done with $\pi_{-\frac{1}{2},0}$ and $\pi_{\frac{1}{2},1}$ in Proposition 2.2.12) so as to let the Heisenberg group show in the picture. However, one can still define a pseudodifferential analysis acting on the space of just one representation, using the one-sheeted hyperboloid as a phase space. Going from the principal series of $SL(2,\mathbb{R})$ to the complementary series by complex continuation (once both representations have been embedded into the full nonunitary principal series), one sees from [34, Theorem 4.2] that the coefficients of the Rankin–Cohen products which occur in the expansion of the composition of symbols certainly depend on the parameter that specifies the representation within its series.

The preceding considerations, to the effect that the anaplectic analyses corresponding to two distinct values of v lead to the same sharp composition of symbols, show that the structure of the composition formula is far from revealing the details of a quantization theory in general. We now show that a certain limit of the v-anaplectic series of analyses contains a part which fits with analysis of the usual kind.

When v is an even integer, it is possible to save most of the v-anaplectic analysis. It would be pleasant to set $v = 0$, but (4.1.1) would then cease to be meaningful: instead, we set $v = -2$, in which case (4.1.1) becomes

$$f_{i,0}(x) = \frac{i}{(2\pi)^{\frac{1}{2}}} \left[f_0(ix) - f_0(-ix) \right],$$

$$f_{i,1}(x) = -\frac{1}{(2\pi)^{\frac{1}{2}}} \left[f_1(ix) + f_1(-ix) \right]. \tag{4.2.10}$$

A function u on the real line lies in $\mathfrak{A}_0 := \mathfrak{A}_{-2}$ if $u = (f_0)_{\text{even}} + (f_1)_{\text{odd}}$ for a pair (f_0, f_1) of *nice* functions satisfying (4.2.10). The map $(f_0, f_1, f_{i,0}, f_{i,1}) \mapsto u$ is still one to one since if f_0 is odd and both f_0 and $f_{i,0}$, linked by (4.2.10), are nice, it follows from the Phragmén–Lindelöf lemma that $f_0 = 0$ [38, p. 6]; the same goes if f_1 is even and both f_0 and $f_{i,0}$ are nice. That not much remains of the anaplectic theory in the case when v is an odd integer is due to the fact that, though one might certainly set $v = -1$ in (4.1.1), the map $f \mapsto u$ would then cease to be one to one, as is easily ascertained: anyway, turning – as is possible – around this difficulty, one does not get anything essentially new in view of Proposition 4.1.2.

We thus assume, from now on, that $v \equiv 0 \bmod 2$ and fix $v = -2$ when dealing with the $\mathbb{C}^4$-realization, though the space $\mathfrak{A}_0$ could of course also be identified with the space $\mathfrak{A}_{-2j}$ defined in the usual way after an arbitrary number $j = 1, 2, \ldots$ has been chosen.

Some special function formulas will help: with the notation of [21], the error functions Erf and Erfc are defined as

$$\text{Erf}(x) = \frac{2}{\sqrt{\pi}} \int_0^x e^{-t^2} \, dt,$$

$$\text{Erfc}(x) = \frac{2}{\sqrt{\pi}} \int_x^\infty e^{-t^2} \, dt = 1 - \text{Erf}(x). \tag{4.2.11}$$

On the other hand, $(H_k)_{k \geq 0}$ is the sequence of Hermite polynomials, characterized by the eigenvalue equation

$$\left[\frac{d^2}{dt^2} + (2k + 1 - t^2) \right] \left(e^{-\frac{t^2}{2}} H_k(t) \right) = 0 \tag{4.2.12}$$

together with the normalizing conditions

$$\int_{-\infty}^\infty e^{-t^2} [H_k(t)]^2 \, dt = 2^k k! \pi^{\frac{1}{2}}, \qquad H_k(+\infty) = +\infty. \tag{4.2.13}$$

From [21, p. 331], the entire function $D_k(2\pi^{\frac{1}{2}}x)$ is, for every $k \in \mathbb{Z}$, a nice function in the sense of Definition 2.2.1 since, as $x \to +\infty$, it is equivalent to a constant times $x^k e^{-\pi x^2}$. For $k \geq 0$, one has (loc.cit., p. 326)

$$D_k(2\pi^{\frac{1}{2}}x) = 2^{-\frac{k}{2}} e^{-\pi x^2} H_k(\sqrt{2\pi}x),$$

$$D_{-k-1}(2\pi^{\frac{1}{2}}x) = \frac{(-1)^k}{k!} \left(\frac{\pi}{2}\right)^{\frac{1}{2}} e^{-\pi x^2} \left(\frac{1}{2\pi^{\frac{1}{2}}} \frac{d}{dx}\right)^k \left[e^{2\pi x^2} \operatorname{Erfc}(\sqrt{2\pi}x)\right]. \quad (4.2.14)$$

In particular,

$$D_0(2\pi^{\frac{1}{2}}x) = e^{-\pi x^2},$$

$$D_1(2\pi^{\frac{1}{2}}x) = 2\pi^{\frac{1}{2}} x e^{-\pi x^2},$$

$$D_{-1}(2\pi^{\frac{1}{2}}x) = \left(\frac{\pi}{2}\right)^{\frac{1}{2}} e^{\pi x^2} \operatorname{Erfc}(\sqrt{2\pi}x),$$

$$D_{-2}(2\pi^{\frac{1}{2}}x) = e^{-\pi x^2} - 2^{\frac{1}{2}} \pi x e^{\pi x^2} \operatorname{Erfc}(\sqrt{2\pi}x). \quad (4.2.15)$$

With the help of Definition 4.1.1 and (4.1.9) and (4.1.10), one finds

$$\psi^0(x) = e^{-\pi x^2},$$

$$\chi^1(x) = 2\pi^{\frac{1}{2}} x e^{-\pi x^2},$$

$$\psi^{-2}(x) = e^{-\pi x^2} + 2^{\frac{1}{2}} \pi x e^{\pi x^2} \operatorname{Erf}(\sqrt{2\pi}x),$$

$$\chi^{-1}(x) = -\left(\frac{\pi}{2}\right)^{\frac{1}{2}} e^{\pi x^2} \operatorname{Erf}(\sqrt{2\pi}x). \quad (4.2.16)$$

It is of course not necessary to restart the v-anaplectic theory from scratch when $v = -2$, since we may regard it as a limiting case: looking at the coefficients which occur on the right-hand sides of (4.1.9) and (4.1.10), one should observe that $\frac{\Gamma(v+2j+1)}{\Gamma(v+1)}$ and $\frac{\Gamma(v+2j+2)}{\Gamma(v+1)}$, as functions of v, are regular at $v = -2$ for every $j \in \mathbb{Z}$: each of them vanishes at this point if and only if $j \geq 1$. One may then apply the definitions of Theorem 4.1.5, obtaining the following.

Theorem 4.2.1. *In the space $\mathfrak{A}_0$, the eigenvalues of the harmonic oscillator $L = \pi(Q^2 + P^2)$ are the numbers $\frac{1}{2} + j$, $j \in \mathbb{Z}$. Every eigenvalue is simple. When $j \geq 0$, the corresponding eigenspace is the same as that obtained in usual analysis, i.e., it is generated by the function $A^{*j} \psi^0$; when $j \leq -1$, the corresponding eigenspace is generated by the function $A^{-j-1} \chi^{-1}$.*

We shall now analyze which part of the v-anaplectic theory subsists in the case under study, at the same time showing that the analysis obtained extends the usual one. Let us emphasize at once that the theory obtained has nothing to do with the direct sum of two parts: one which would correspond to some space containing the usual Hermite functions and another one which would correspond to the negative part (based on the use of some version of the error function) of the spectrum of the harmonic oscillator. For, as will be seen, one can define the first space so as to make it invariant both under the Heisenberg representation and the 0-anaplectic representation. But nothing comparable can occur on the other side since, starting from any (anaplectic) eigenfunction of the harmonic oscillator corresponding to a negative eigenvalue and applying a suitable power of the raising operator $A^* = \pi^{\frac{1}{2}} (Q - iP)$, one can reach usual Hermite functions. Indeed, note that (4.1.11) are still valid, whether one substitutes 0 or -2 (or any even integer) for ν.

There is nothing wrong with using the value $\nu = -2$ so far as the linear form Int is concerned, and we may use the results of Sect. 4.1.

Proposition 4.2.2. *In 0-anaplectic (this is by definition the same as (-2)-anaplectic) analysis, the linear form* Int *vanishes on odd functions. It coincides with the linear form* $u \mapsto e^{\frac{i\pi}{4}} \int_{-\infty}^{\infty} u(x)\,dx$ *on the subspace of* $\mathfrak{A}_0$ *generated by the usual Hermite functions* ψ^{2j}, $j \geq 0$. *The functions* ψ^{-2j-2}, $j \geq 0$, *are not integrable on the real line, but one has the equation*

$$\mathrm{Int}\,[\psi^{-2j-2}] = \frac{(-1)^{j+1}\,e^{\frac{i\pi}{4}}}{1.3\ldots(2j+1)}. \tag{4.2.17}$$

Proof. That Int vanishes on odd functions is a consequence of Definition 8.6 (applied with $\nu = -2$) together with the fact that the components f_0, $f_{i,0}$ of the $\mathbb{C}^4$-realization of an odd function are zero. Since the ν-anaplectic representation combines with the Heisenberg representation in the usual way, one has, denoting $\mathcal{F}_{\mathrm{ana}}^{\nu}$ as $\mathcal{F}_{\mathrm{ana}}^{(0)}$ when $\nu \equiv 0 \bmod 2$, $\mathcal{F}_{\mathrm{ana}}^{(0)} A^* = -iA^* \mathcal{F}_{\mathrm{ana}}^{(0)}$: since, for $j \geq 0$, $\psi^{2j} = (A^*)^{2j+2} \psi^{-2}$, and, as displayed in (4.1.24), $\mathcal{F}_{\mathrm{ana}}^{(0)} \psi^{-2} = -e^{\frac{i\pi}{4}} \psi^{-2}$, one obtains

$$\mathcal{F}_{\mathrm{ana}}^{(0)} \psi^{2j} = (-iA^*)^{2j+2} \mathcal{F}_{\mathrm{ana}}^{(0)} \psi^{-2} = (-1)^j e^{\frac{i\pi}{4}} (A^*)^{2j+2} \psi^{-2} = (-1)^j e^{\frac{i\pi}{4}} \psi^{2j} : \tag{4.2.18}$$

consequently,

$$\mathrm{Int}\,[\psi^{2j}] = \left(\mathcal{F}_{\mathrm{ana}}^{(0)} \psi^{2j} \right)(0) = (-1)^j e^{\frac{i\pi}{4}} \psi^{2j}(0). \tag{4.2.19}$$

The same calculation, using the usual integral on the line instead of the linear form Int and the usual Fourier transformation in place of $\mathcal{F}_{\mathrm{ana}}^{(0)}$ leads to the same result, save for the factor $e^{\frac{i\pi}{4}}$.

The functions ψ^{-2j-2} can never be integrable on the line, since they are formal eigenfunctions of the harmonic oscillator corresponding to negative eigenvalues. Still, using (4.1.11), one can write

$$\psi^{-2j-2} = \frac{1}{(2j+1)!} A^{2j} \psi^{-2}, \qquad j \geq 1, \tag{4.2.20}$$

from which one obtains

$$\mathcal{F}_{\text{ana}}^{(0)}\,\psi^{-2j-2} = \frac{1}{(2j+1)!}\,(iA)^{2j}\,\mathcal{F}_{\text{ana}}^{(0)}\,\psi^{-2}$$

$$= \frac{(-1)^j}{(2j+1)!}\left(-e^{\frac{i\pi}{4}}A^{2j}\,\psi^{-2}\right) = (-1)^{j+1}\,e^{\frac{i\pi}{4}}\,\psi^{-2j-2}, \quad (4.2.21)$$

so that

$$\text{Int}\,[\psi^{-2j-2}] = (-1)^{j+1}\,e^{\frac{i\pi}{4}}\,\psi^{-2j-2}(0); \quad (4.2.22)$$

finally, one has [21, p. 324] $D_{-2j-2}(0) = (1.3.\dots(2j+1))^{-1}$, and ψ^{-2j-2} is by definition the even part of the function $x \mapsto D_{-2j-2}(2\pi^{\frac{1}{2}}x)$. $\square$

We now turn to the consideration of the 0-anaplectic representation Ana_0. To take benefit from the fact that the case when $\nu = -2$ is a limit of cases where Theorem 4.1.8 is already known to apply, we must first interpret Ana_0 as a representation of the universal cover $G^{(\infty)}$ of G. Consider the linear space E generated by functions $u(x) = p(x)\exp(-\pi q(x))$, where p and q are complex polynomials, q is of degree 2 and has a top-order coefficient with a positive real part: it contains the Hermite functions. On the other hand, for any $u \in E$, the vector $(u_{\text{even}}, u_{\text{odd}}, 0, 0)$ is a $\mathbb{C}^4$-realization of u, so that $E \subset \mathfrak{A}_0$: note that the simple recipe just indicated toward the construction of a $\mathbb{C}^4$-realization only works when $\nu \equiv 0 \mod 2$. We now consider the effect on functions in E of operators from the representation Ana_0 or from the metaplectic representation $\text{Met}^{(1)}$.

Proposition 4.2.3. *On the space E defined as the smallest linear space of functions on the real line containing the standard Gaussian function and stable under transformations $u \mapsto u_1$ with $u_1(x) = u(ax)$, $a > 0$ or $u_1(x) = u(x)\,e^{i\pi cx^2}$, $c \in \mathbb{R}$, as well as under the operators Q and P, the representation Ana_0 agrees with the metaplectic representation.*

Proof. Again, we shall use generators. On matrices g such as $\begin{pmatrix} 1 & 0 \\ c & 1 \end{pmatrix}$ or $\begin{pmatrix} a & 0 \\ 0 & a^{-1} \end{pmatrix}$ (identified with elements of $G^{(\infty)}$), $\text{Met}^{(1)}(g)$ and $\text{Ana}_0(g)$ coincide. Consider now the element $g = \exp\frac{\pi}{2}\begin{pmatrix} 0 & 1 \\ -1 & 0 \end{pmatrix} \in G^{(\infty)}$ introduced in Theorem 4.1.8, and denote by the same letter its canonical image in $G^{(2)} = \widetilde{\text{Sp}}(1,\mathbb{R})$: according to the definitions recalled in Sect. 2.1, one has $\text{Met}^{(1)}(g) = e^{-\frac{i\pi}{4}}\mathcal{F}$ whereas, from Theorem 4.1.8, $\text{Ana}_0(g) = e^{-\frac{i\pi}{2}}\mathcal{F}_{\text{ana}}^0$. Since ψ^0, the standard Gaussian function, is invariant under $\mathcal{F}$ but, according to (4.1.24), is multiplied by $e^{\frac{i\pi}{4}}$ under $\mathcal{F}_{\text{ana}}^0$, one sees that ψ^0 has the same images under $\text{Met}^{(1)}(g)$ and under $\text{Ana}_0(g)$: as the operator A^* undergoes the same transformation under adjunction by $\mathcal{F}$ or by $\mathcal{F}_{\text{ana}}^0$, one sees that $\text{Met}^{(1)}(g)$ and $\text{Ana}_0(g)$ agree on the space generated by Hermite functions and, as an easy extension, on E. $\square$

Of course, the novelty of the 0-anaplectic representation is that it makes sense on a space much larger than E. When $\nu \equiv 0 \mod 2$, it is no longer possible to define an invariant nondegenerate pseudoscalar product. Indeed, the coefficient $\frac{\Gamma(\nu+1)}{\Gamma(-\nu)}$ which occurs on the right-hand side of (4.1.20) is infinite when $\nu = -2$. Note that, in the

case when $u \in E$, one has $f_{i,0} = f_{i,1} = 0$ as already mentioned, and $(u \mid u)$, reduced to
the sum of its first two terms, is then well defined and coincides with $2^{-\frac{1}{2}} \| u \|_{L^2(\mathbb{R})}^2$.
In another direction, for every $u \in \mathfrak{A}_0$, one may define instead $((u \mid u))$ as the "infinite
part" of (4.1.20), i.e., as

$$((u \mid u)) = 2^{\frac{1}{2}} \int_0^\infty (|f_{i,0}|^2 - |f_{i,1}|^2) \, dx. \tag{4.2.23}$$

The pseudoscalar product so defined is still invariant both under the 0-anaplectic
representation and under the Heisenberg representation. The eigenstates of
the harmonic oscillator are still pairwise orthogonal. After having multiplied
by $\frac{\Gamma(-\nu)}{\Gamma(\nu+1)}$ the right-hand sides of (4.1.23) which led to the normalization
used in Theorem 4.1.9, one obtains $((\psi^0 \mid \psi^0)) = ((\psi^2 \mid \psi^2)) = \cdots = 0$ and
$((\chi^1 \mid \chi^1)) = ((\chi^3 \mid \chi^3)) = \cdots = 0$, i.e., the pseudoscalar product under consideration vanishes when considered on Hermite functions: it also vanishes when
considered on a pair of functions, at least one of which lies in E. On the other hand,
one verifies from the same calculation that

$$((\psi^{-2k-2} \mid \psi^{-2k-2})) = \frac{1}{2(2k+1)!}, \qquad ((\chi^{-2k-1} \mid \chi^{-2k-1})) = -\frac{1}{2(2k)!} \tag{4.2.24}$$

if $k = 0, 1, \ldots$.

In ν-analysis with $\nu \in \mathbb{C} \backslash \mathbb{Z}$, one obtains from (4.1.11) that

$$(Q+iP)^{-1} \chi^{\nu+2j-1} = \frac{\pi^{\frac{1}{2}}}{\nu+2j} \, \psi^{\nu+2j},$$

$$(Q+iP)^{-1} \psi^{\nu+2j} = \frac{\pi^{\frac{1}{2}}}{\nu+2j+1} \chi^{\nu+2j+1}. \tag{4.2.25}$$

In the case when $\nu \equiv 0 \mod 2$, $(Q+iP)^{-1}$ ceases to be well defined since the
operator $Q+iP$ is neither onto nor one to one as an endomorphism of $\mathfrak{A}_0$. When
attention is restricted to usual analysis, one can of course define a *right inverse*
$(Q+iP)^{-1}$ of this operator by the equations

$$(Q+iP)^{-1} \chi^{2j-1} = \frac{\pi^{\frac{1}{2}}}{2j} \, \psi^{2j}, \qquad\qquad j \geq 1,$$

$$(Q+iP)^{-1} \psi^{2j} = \frac{\pi^{\frac{1}{2}}}{2j+1} \chi^{2j+1}, \qquad\qquad j \geq 0: \tag{4.2.26}$$

these equations define $(Q+iP)^{-1}$ on the linear space generated by Hermite func-
tions, and the operator can be extended to $\mathcal{S}(\mathbb{R})$ or to $L^2(\mathbb{R})$ by means of the equation

$$A^{-1} := A^* (L + \frac{1}{2})^{-1} = (L - \frac{1}{2})^{-1} A^*. \tag{4.2.27}$$

However, it is not possible to base on the use of such an operator (and of its conjugates under operators of the metaplectic representation) an ascending pseudodifferential analysis, by a generalization of Definition 3.1.2. What goes wrong is that the analogue of Lemma 3.1.5 does not hold any more: equations (3.1.20) still hold, but only for nonnegative exponents.

Chapter 5
Pseudodifferential Analysis and Modular Forms

This chapter is meant as another motivation (cf. introduction) for the construction of the ascending pseudodifferential analysis, rather than as an introduction to a new point of view in modular form theory: we hope to come back to possible developments in this direction at some later occasion. We here wish to show that the parallel treatments of anaplectic analysis and associated alternative pseudodifferential analysis on one hand, of usual analysis and pseudodifferential analysis on the other hand, extend, up to some point, as parallel sources of holomorphic modular forms on one hand, nonholomorphic modular forms on the other hand. In particular, let us direct the interested reader to Remark 5.1.1 at the end of this section, which puts on an absolutely equal footing the Rankin–Cohen brackets of holomorphic modular forms, well known to modular form theorists, and their analogues in nonholomorphic modular form theory, which do not seem to have kept their attention. As explained in the introduction, making this possible was one of the initial aims of the present work.

Since the question is a very natural one, we shall explain, in Sect. 5.2, why Rankin–Cohen brackets have no role to play in usual pseudodifferential analysis and which superficially similar, but distinct, brackets do.

5.1 The Eisenstein, Theta, Poincaré, and Alternative Poincaré Distributions

In the usual point of view [36], automorphic distributions in the plane are tempered distributions invariant under the linear action on $\mathbb{R}^2$ of some arithmetic group, say $SL(2, \mathbb{Z})$. For instance, such is the Dirac comb

$$\mathfrak{D} = 2\pi \sum_{\substack{(j,k) \in \mathbb{Z} \times \mathbb{Z} \\ |j| + |k| \neq 0}} \delta(x_1 - j)\, \delta(x_2 - k), \tag{5.1.1}$$

which can be decomposed as an integral $\mathfrak{D} = 2\pi + \int_{-\infty}^{\infty} \mathfrak{E}_{i\lambda}^{\sharp} \, d\lambda$, where the "Eisenstein" distribution $\mathfrak{E}_{i\lambda}^{\sharp}$ is homogeneous of degree $-1 - i\lambda$. Of course, the adjective "usual" must be taken here only in the sense of "nonalternative."

This extends without modification [37] to the higher-dimensional case, but there is a special feature in the two-dimensional case. There are two families (u_z^0) and (u_z^1) of (respectively, even and odd) functions in $S(\mathbb{R})$, parametrized by $z \in \Pi$, such that if $\mathfrak{S}$ is an automorphic distribution and if Op is the Weyl calculus, the functions $z \mapsto f_j(z) = (u_z^j \,|\, \text{Op}(\mathfrak{S}) \, u_z^j)$ are automorphic: moreover, if $\mathfrak{S}$ is homogeneous of degree $-1 - i\lambda$, its pair of images (which characterizes $\mathfrak{S}$) consists of generalized eigenfunctions of the hyperbolic Laplacian, for the eigenvalue $\frac{1+\lambda^2}{4}$; in other words, these two functions are nonholomorphic modular forms, making up a set of Cauchy data for the automorphic wave equation in the sense of Lax–Phillips [18], as shown in [35, Sect. 18]. There is a natural automorphic distribution $\mathfrak{B}$, the decomposition of which contains *all* homogeneous automorphic distributions: we called it the "Bezout distribution" in [36] but we shall change its name to that of "Poincaré distribution."

The map $\mathfrak{S} \mapsto (f_0, f_1)$ establishes a dictionary between modular distributions (i.e., homogeneous automorphic distributions) and pairs of nonholomorphic modular forms. The notion of modular distribution is slightly more precise than that of nonholomorphic modular form: for instance, the Eisenstein series $E_{\frac{1 \pm i\lambda}{2}}$ are proportional, whereas $\mathfrak{E}_{i\lambda}^{\sharp}$ and $\mathfrak{E}_{-i\lambda}^{\sharp}$ are related by the *symplectic* Fourier transformation on $\mathbb{R}^2$ (the one with the integral kernel $(x,y) \mapsto \exp(2i\pi(x_1 y_2 - x_2 y_1))$); something similar holds in relation to the correspondence from "cusp distributions" to nonholomorphic cusp forms.

This completes the part of the "usual" theory for which we plan to obtain a alternative substitute in this section, starting from Proposition 5.1.1 which, just as in the usual case, makes it possible to recover a symbol from a special *diagonal* set of matrix elements of the associated operator against the family $(\phi_z^{m+1})_{z \in \Pi}$: "diagonal" means that we only consider pairs of functions ϕ_z^j with the same z, not the same j.

Proposition 5.1.1. *Let* $h = h_m \in L_m^2(\mathbb{R}^2)$. *For every* $z \in \Pi$, *one has*

$$(\phi_z^{m+1} \,|\, \text{Op}_m^{\text{asc}}(h_m) \, \phi_z) = \pi^{\frac{m+1}{2}} \, (\Theta_m h_m)(z). \tag{5.1.2}$$

Proof. From (3.1.11), one has

$$(\phi_z^{m+1} \,|\, \text{Op}_m^{\text{asc}}(h_m) \, \phi_z)$$
$$= \frac{m}{4\pi} \pi^{\frac{m+1}{2}} \int_{\Pi} (\Theta_m h)(w) \, (\phi_z^{m+1} \,|\, A_w^{-m-1} \, \phi_z) \, (\text{Im } w)^{m+1} \, d\mu(w) \tag{5.1.3}$$

and from Lemma 3.2.1, one has

$$(\phi_z^{m+1} \,|\, A_w^{-m-1} \, \phi_z) = (-2i)^{m+1} \, (\text{Im } z)^{\frac{m+1}{2}} \, (\bar{w} - z)^{-m-1}, \tag{5.1.4}$$

so that

$$
\begin{aligned}
&(\phi_z^{m+1} \,|\, \mathrm{Op}_m^{\mathrm{asc}}(h_m)\,\phi_z) \\
&= \frac{m}{4\pi}\,\pi^{\frac{m+1}{2}}\,(\mathrm{Im}\,z)^{\frac{m+1}{2}} \int_{\Pi} (\Theta_m h)(w) \left[\frac{i}{2}\,(\bar{w}-z)\right]^{-m-1} (\mathrm{Im}\,w)^{m+1}\,d\mu(w),
\end{aligned}
$$
(5.1.5)

which reduces to the right-hand side of (5.1.2) if one uses the Bergman's reproducing kernel of the Hilbert space $\mathcal{H}_{m+1}$, as defined in Proposition 2.1.1.   □

Interesting symbols, from the point of view of arithmetic, do not generally lie in $L_m^2(\mathbb{R}^2)$, not even – at least for small $m \geq 1$ – in the space $(S'_{\mathrm{weak}}(\mathbb{R}^2))^{\uparrow}$ introduced before Lemma 3.2.4. Hence, they do not fit within the range of applicability of Proposition 3.2.5. Still, as shown by Proposition 5.1.1, there remains the question whether $\Theta_m h_m$ can be defined for such a function, or rather distribution, h_m: we answer it in the affirmative in a few cases, so as to show the link between "alternative automorphic distributions" in the plane (denoted as $\mathfrak{S}$ rather than h so as to recall their singular nature) and sequences of holomorphic modular forms of weights $2, 3, \ldots$.

Any function $h \in S(\mathbb{R}^2)$ can be decomposed as $h = \sum_{m \in \mathbb{Z}} h_m$, where h_m satisfies (2.1.8). By duality, this defines a decomposition of tempered distributions into isotypic components as well, setting $\langle \mathfrak{S}_m, h \rangle = \langle \mathfrak{S}, h_{-m} \rangle$. In polar coordinates, with $x_1 + ix_2 = r e^{i\theta}$, one has

$$
\begin{aligned}
h_{-m}(x) &= \phi_{-m}(r)\,e^{im\theta} \\
&= \frac{1}{2\pi} \left(\frac{x_1 + ix_2}{|x|}\right)^m \int_0^{2\pi} h(|x|\cos\omega, |x|\sin\omega)\,e^{-im\omega}\,d\omega,
\end{aligned}
$$
(5.1.6)

a formula to be used in a moment.

Consider first the "complete" Dirac comb $\mathfrak{S} = \mathfrak{D}^{\bullet}$ obtained by adding 2π times the Dirac mass at the origin to the distribution $\mathfrak{D}$. Besides being invariant under the linear changes of coordinates associated with elements of $SL(2,\mathbb{Z})$, it is an automorphic object on a second account too: it is invariant under the *Euclidean* Fourier transformation $\mathcal{F}$ as well as under the multiplication by the function $\exp(2i\pi\,(x_1^2 + x_2^2))$. In other words, looking back at (2.1.6), it is invariant under the transformations $\mathrm{Met}^{(2)}(g)$ whenever g lies in the subgroup Γ_2 of $SL(2,\mathbb{Z})$ generated by the matrices $\begin{pmatrix} 0 & 1 \\ -1 & 0 \end{pmatrix}$ and $\begin{pmatrix} 1 & 2 \\ 0 & 1 \end{pmatrix}$.

Note that Γ_2 coincides with the subgroup Γ_2' of $SL(2,\mathbb{Z})$ consisting of matrices $g = \begin{pmatrix} a & b \\ c & d \end{pmatrix}$, such that $\begin{pmatrix} a & b \\ c & d \end{pmatrix} \equiv \begin{pmatrix} 0 & 1 \\ 1 & 0 \end{pmatrix} \bmod 2$ or $\begin{pmatrix} a & b \\ c & d \end{pmatrix} \equiv \begin{pmatrix} 1 & 0 \\ 0 & 1 \end{pmatrix} \bmod 2$, as the following argument shows. First, it is immediate that $\Gamma_2 \subset \Gamma_2'$. Call a pair $\{a, c\}$ of integers admissible if $(a, c) = 1$ and a or c is even. Given such a pair with a odd, an assumption which does not diminish the generality since $\begin{pmatrix} 0 & 1 \\ -1 & 0 \end{pmatrix} \begin{pmatrix} a & : \\ c & : \end{pmatrix} = \begin{pmatrix} c & : \\ -a & : \end{pmatrix}$, choosing d such that $ad \equiv 1 \bmod 2c$, one obtains a matrix $g = \begin{pmatrix} a & b \\ c & d \end{pmatrix} \in \Gamma_2'$. If $g' = \begin{pmatrix} a & b' \\ c & d' \end{pmatrix} \in \Gamma_2'$ as well, d and d' have the same parity (opposed to that of c) so that $2 | d - d'$; on the other hand, since $ad' - b'c = ad - bc$, or $a(d' - d) = c(b' - b)$ and $(a, c) = 1$, one has

$a|b' - b$; if a is odd, $b' - b$ is divisible by a and by 2, hence by $2a$; if a is even, c is odd, hence relatively prime with $2a$, so that $2a$, which divides $a(d' - d) = c(b' - b)$, must divide $b' - b$. Finally, one always has $b' - b = 2ja$ for some $j \in \mathbb{Z}$, which implies $d' - d = 2jc$ and $g' = \left(\begin{smallmatrix} a & b+2ja \\ c & d+2jc \end{smallmatrix} \right) = \left(\begin{smallmatrix} a & b \\ c & d \end{smallmatrix} \right) \left(\begin{smallmatrix} 1 & 2j \\ 0 & 1 \end{smallmatrix} \right)$: this shows that, given $g = \left(\begin{smallmatrix} a & b \\ c & d \end{smallmatrix} \right) \in \Gamma'_2$, whether $g \in \Gamma_2$ only depends on the pair $\{a, c\}$.

Introduce the (temporary) notion that two admissible pairs $\{a, c\}$ and $\{a', c'\}$ are equivalent if, when they are completed as the first columns of two matrices g and g' in Γ'_2, either both g and g' lie in Γ_2 or neither one nor the other does. Then, since $\left(\begin{smallmatrix} 1 & 0 \\ \pm 2 & 1 \end{smallmatrix} \right) \left(\begin{smallmatrix} a & : \\ c & : \end{smallmatrix} \right) = \left(\begin{smallmatrix} a & : \\ c \pm 2a & : \end{smallmatrix} \right)$, the pairs $\{a, c\}$ and $\{a, c \pm 2a\}$ are equivalent; since $\{a, c\}$ and $\{c, -a\}$ are equivalent, the pair $\{a, c\}$ is also equivalent to $\{a \pm 2c, c\}$. Let $\{a, c\}$ be an admissible pair such that $ac \neq 0$, so that $0 < |a| < |c|$ or $0 < |c| < |a|$: in the first case, one has either $|c + 2a| < |c|$ or $|c - 2a| < |c|$; in the second one, one has either $|a + 2c| < |a|$ or $|a - 2c| < |a|$. Finally, the pair $\{a, c\}$ is equivalent to a pair $\{a', c'\}$ with $|a'| + |c'| < |a| + |c|$, and by induction to a pair $\{a', c'\}$ with $a'c' = 0$, of necessity of the kind $\{\pm 1, 0\}$ or $\{0, \pm 1\}$. Completing such a pair to a matrix in Γ'_2 of course yields a matrix in Γ_2, so that $\Gamma_2 = \Gamma'_2$.

Denoting as H the group generated by the matrix $\left(\begin{smallmatrix} 1 & -2 \\ 0 & 1 \end{smallmatrix} \right)$, the class of a matrix $\left(\begin{smallmatrix} a & b \\ c & d \end{smallmatrix} \right) \in \Gamma_2$ in Γ_2/H is characterized by the admissible pair $\{a, c\}$. Also, recalling that the Hecke group traditionally denoted as $\Gamma_0(2)$ consists of all matrices $g = \left(\begin{smallmatrix} a & b \\ c & d \end{smallmatrix} \right)$ such that c is even, note that the conjugation map $g \mapsto \left(\begin{smallmatrix} 1 & 0 \\ 1 & 1 \end{smallmatrix} \right) g \left(\begin{smallmatrix} 1 & 0 \\ -1 & 1 \end{smallmatrix} \right)$ is an isomorphism from $\Gamma_0(2)$ onto Γ_2.

With the help of (5.1.6), it is immediate to obtain the isotypic components of $\mathfrak{D}^\bullet$ as follows:

$$\langle (\mathfrak{D}^\bullet)_0, h \rangle = h(0) + \sum_{n \geq 1} \mathrm{Sq}_2(n) \, \phi_0(\sqrt{n}),$$

$$\langle (\mathfrak{D}^\bullet)_m, h \rangle = \sum_{n \geq 1} \mathrm{Sq}_{2,m}(n) \, \phi_{-m}(\sqrt{n}) \quad \text{if } m \neq 0, \tag{5.1.7}$$

where $\mathrm{Sq}_2(n)$ denotes the number of decompositions of n as a sum of two squares (of any sign; the order is taken into account) and, for $m \geq 1$,

$$\mathrm{Sq}_{2,m}(n) = \sum_{\substack{(j,k) \in \mathbb{Z} \times \mathbb{Z} \\ j^2 + k^2 = n}} \left(\frac{j + ik}{\sqrt{j^2 + k^2}} \right)^m. \tag{5.1.8}$$

Even though $(\mathfrak{D}^\bullet)_m$ is a distribution, rather than an L^2-function, one may still, assuming $m \geq 0$, use (2.1.9) in the form

$$(\Theta_m (\mathfrak{D}^\bullet)_m)(z) = z^{-m-1} \langle (\mathfrak{D}^\bullet)_m, x \mapsto (x_1 + ix_2)^m e^{-\frac{i\pi}{z}|x|^2} \rangle : \tag{5.1.9}$$

from (5.1.7), one obtains

$$(\Theta_m (\mathfrak{D}^\bullet)_m)(z) = z^{-m-1} \sum_{n \geq 1} n^{\frac{m}{2}} \mathrm{Sq}_{2,m}(n) \, e^{-\frac{i\pi n}{z}}. \tag{5.1.10}$$

As a consequence of property (ii) from Proposition 2.1.1, and of the invariance of $(\mathfrak{D}^{\bullet})_m$ under the two-dimensional Fourier transformation, the function $\Theta_m (\mathfrak{D}^{\bullet})_m$ is invariant under (the extension of) the operator $i\mathcal{D}_{m+1} \left(\begin{smallmatrix} 0 & 1 \\ -1 & 0 \end{smallmatrix}\right)$: hence,

$$(\Theta_m (\mathfrak{D}^{\bullet})_m)(z) = (-1)^m i \sum_{n \geq 1} n^{\frac{m}{2}} \operatorname{Sq}_{2,m}(n) \, e^{i\pi n z}. \tag{5.1.11}$$

The function $f(z) = \sum_{n \geq 1} n^{\frac{m}{2}} \operatorname{Sq}_{2,m}(n) \, e^{i\pi n z}$ is thus a modular form of weight $m + 1$ for the group Γ_2 and for an appropriate multiplier, since it satisfies the equation

$$f(z) = \kappa \left(\frac{z}{i}\right)^{-m-1} f\left(-\frac{1}{z}\right) \tag{5.1.12}$$

with $\kappa = (-i)^m$: actually, m is even if $f \neq 0$. Note that the functional equation (5.1.12), together with a Fourier expansion as in (5.1.11), characterizes the class of modular forms considered in the first section of [23].

Since the function f obtained from the distribution $(\mathfrak{D}^{\bullet})_m$ is a theta series, it is natural to call this distribution a theta distribution. We will not say more about it here, except for pointing at one way of writing the fact that this distribution, an isotypic component of $\mathfrak{D}^{\bullet}$, must also be invariant under the Fourier transformation. If $h \in L^2_{-m}(\mathbb{R}^2) \cap S(\mathbb{R}^2)$ with $m \geq 0$, $h(x) = \phi(|x|) \left(\frac{x_1 + i x_2}{|x|}\right)^m$, and if $\mathcal{F}h$ has a similar expression, with ϕ replaced by ψ, let us set

$$\phi(|x|) = \Phi(|x|^2), \qquad \psi(|\xi|) = \Psi(|\xi|^2): \tag{5.1.13}$$

the Bochner–Hecke formula (2.1.7) expresses itself as

$$\Psi(s^2) = 2\pi (-i)^m \int_0^{\infty} t \, \Phi(t^2) \, J_m(2\pi s t) \, dt. \tag{5.1.14}$$

The invariance of the distribution $(\mathfrak{D}^{\bullet})_m$ under the Fourier transformation is then equivalent to the identity

$$\sum_{n \geq 1} \operatorname{Sq}_{2,m}(n) \, \Phi(n) = \sum_{n \geq 1} \operatorname{Sq}_{2,m}(n) \, \Psi(n) \tag{5.1.15}$$

in the case when $m \geq 1$ and to the identity

$$\Phi(0) + \sum_{n \geq 1} \operatorname{Sq}_2(n) \, \Phi(n) = \Psi(0) + \sum_{n \geq 1} \operatorname{Sq}_2(n) \, \Psi(n) \tag{5.1.16}$$

in the case when $m = 0$. This last identity is exactly (except for some easy possible relaxing of the technical assumptions about ϕ, ψ) the version of Voronoi's formula [40] given in [12, 37]: note that the case when Φ is the characteristic function of some interval $[0, T]$ was emphasized in the first two references. Recall that Voronoi's formula is a standard introduction to Gauss' circle problem (the search for a good estimate of the error term in asymptotics of the number of points with integral coordinates in a circle with large radius).

Many identities, in some ways similar to Voronoi's identity, are to be found in [37]. They are based on the decomposition of certain automorphic distributions as continuous superpositions of Eisenstein distributions: things are more interesting there than in the present context, since modular forms, of holomorphic and nonholomorphic type, and L-functions also play a role.

We now introduce the Poincaré and alternative Poincaré distributions, the first of which was introduced in [36] under the name of "Bezout" distribution, and denoted as $\mathfrak{B}$. Our interest in it came from the fact that it is a generating object for all automorphic distributions (loc.cit., p. 34): moreover, the sharp composition of two such objects could then be expressed as the image of $\mathfrak{B}$ under some operator built from the Hecke operators. So far as the construction of $\mathfrak{B}$ is concerned, the familiar idea (only transposed to a distribution setting) is to start from a distribution having already *some* invariance: in that case, the application toward automorphic pseudodifferential analysis we had in mind led us to choosing the distribution $\mathfrak{b}(x_1, x_2) = e^{2i\pi x_1} \delta(x_2 - 1)$, already invariant under the quasiregular action of matrices $\begin{pmatrix} 1 & b \\ 0 & 1 \end{pmatrix}$ with $b \in \mathbb{Z}$. Then, building an automorphic distribution from $\mathfrak{b}$ only requires that one should perform a summation with respect to the set of classes of $SL(2, \mathbb{Z})$ modulo the subgroup just introduced.

We shall do something analogous in alternative analysis, starting this time from a measure μ carried by the circle of radius $N^{\frac{1}{2}}$, $N = 1, 2, \ldots$, to wit, in terms of the coordinate θ such that $N^{\frac{1}{2}} e^{i\theta}$ lies on the circle, a measure $\nu(\theta) \frac{d\theta}{2\pi}$, where the density can be expressed as a Fourier series:

$$\nu(\theta) = \sum_{m \geq 0} \alpha_m e^{-im\theta}. \tag{5.1.17}$$

With this choice of radii, the group Γ_2 comes into the picture (if N is even, the full modular group $SL(2, \mathbb{Z})$ will do instead). Indeed, the Fourier transform h of the measure μ is already invariant under the transformation $\mathcal{F}(e^{2i\pi |x|^2}) \mathcal{F}^{-1}$, where the operator in the middle stands for the multiplication by the function indicated: since $\begin{pmatrix} 0 & 1 \\ -1 & 0 \end{pmatrix} \begin{pmatrix} 1 & 0 \\ 2 & 1 \end{pmatrix} \begin{pmatrix} 0 & -1 \\ 1 & 0 \end{pmatrix} = \begin{pmatrix} 1 & -2 \\ 0 & 1 \end{pmatrix}$, the product of all three operators is the metaplectic operator $\mathrm{Met}^{(2)} \left(\begin{pmatrix} 1 & -2 \\ 0 & 1 \end{pmatrix} \right)$. In other words, h is invariant under all transformations $\mathrm{Met}^{(2)}(\gamma)$ with γ in the group H introduced just before (5.1.7).

One has

$$h(x) = \frac{1}{2\pi} \int_{-\pi}^{\pi} \sum_{m \geq 0} \alpha_m e^{-im\theta} e^{-2i\pi N^{\frac{1}{2}} (x_1 \cos\theta + x_2 \sin\theta)} \, d\theta$$

$$= \sum_{m \geq 0} (-i)^m \alpha_m h_m(x), \tag{5.1.18}$$

with

$$h_m(x) = \left(\frac{x_1 - ix_2}{|x|} \right)^m J_{|m|}(2\pi N^{\frac{1}{2}} |x|). \tag{5.1.19}$$

Let κ be either the trivial character of Γ_2 or the character trivial on H such that $\kappa \left(\begin{pmatrix} 0 & 1 \\ -1 & 0 \end{pmatrix} \right) = -1$: in the latter case, $\kappa \left(\begin{pmatrix} a & b \\ c & d \end{pmatrix} \right) = (-1)^c$, so that $\kappa \left(\begin{pmatrix} a & b \\ c & d \end{pmatrix} \right)$ can be written as $\kappa(c)$ in both cases. If the series

$$\mathfrak{C} = \sum_{\gamma \in \Gamma_2/H} \kappa(\gamma) \, \mathrm{Met}^{(2)}(\gamma) \, h \tag{5.1.20}$$

can be made to converge, it will define an automorphic object of some kind, relative to the arithmetic group Γ_2 and to the character κ, in view of the identity

$$\mathrm{Met}^{(2)}(\gamma_1) \, \mathfrak{C} = \kappa(\gamma_1)^{-1} \mathfrak{C}, \qquad \gamma_1 \in \Gamma_2 : \tag{5.1.21}$$

note that the defining series follows the familiar Poincaré-type process, considered for distributions. Since, eventually, we are going to decompose the result according to the action of the rotation group, we may as well start from the decomposition (5.1.18) and take advantage of Proposition 2.1.1.

With the notation from this proposition, one has if $m \geq 0$, using [21, p. 93],

$$N^{\frac{m+2}{2}} (\Theta_m h_m)(z) = z^{-m-1} \int_{\mathbb{R}^2} |x|^m J_m(2\pi N^{\frac{1}{2}} |x|) \, e^{-\frac{i\pi}{z} |x|^2} \, dx$$
$$= (-i)^{m+1} e^{i\pi N z}. \tag{5.1.22}$$

For $m < 0$, the formula is slightly more complicated and may be found with the help of [11, p. 103]: assuming that $N = 1$ for simplicity,

$$(\Theta_m h_m)(z) = \frac{1}{2i} \frac{\pi^{|m|-1}}{(|m|-1)!} z^{|m|-1} [1 - \frac{1}{i\pi} \frac{d}{dz}]^{|m|-1} \left(\frac{1 - e^{i\pi z}}{i\pi z} \right). \tag{5.1.23}$$

Specializing to the first case (the only one, incidentally, to be considered in relation with the ascending calculus) and using Proposition 2.1.1, one obtains

$$N^{\frac{m+2}{2}} \Theta_m \left(\sum_{\gamma \in \Gamma_2/H} \kappa(\gamma) \, \mathrm{Met}^{(2)}(\gamma) \, h_m \right)(z) = (-i)^{m+1} \sum_{\gamma \in \Gamma_2/H} \kappa(\gamma) \, \mathcal{D}_{m+1}(\gamma) \, (z \mapsto e^{i\pi N z})$$
$$= (-i)^{m+1} \sum_{\substack{(a,c)=1 \\ ac \equiv 0 \bmod 2}} \kappa(c) \, (-cz+a)^{-m-1} \exp \left(i\pi N \frac{dz-b}{-cz+a} \right). \tag{5.1.24}$$

The exponential factor does not depend on the second column of the matrix $\begin{pmatrix} a & b \\ c & d \end{pmatrix} \in \Gamma_2$, and the last series is convergent provided that $m \geq 2$: the sum is a completely standard Poincaré series. It is a modular form of weight $m+1$ for the group Γ_2 and character κ [4], actually a cusp form: of course, nothing is obtained unless $\kappa(-I) = (-1)^{m+1}$.

When $N = 0$, still assuming $m \geq 2$, one sees that the right-hand side of (5.1.19) vanishes but, if first multiplied by $2^m m! (2\pi N^{\frac{1}{2}})^{-m}$, leads to the function $h_m(x) = (x_1 - i x_2)^m$. Then, $(\Theta_m h_m)(z) = \pi m! i^{-m-1}$, and the left-hand side of (5.1.24) reduces to

$$\pi m! i^{-m-1} \sum_{\substack{(a,c)=1 \\ ac \equiv 0 \bmod 2}} \kappa(c) \, (-cz+a)^{-m-1}, \tag{5.1.25}$$

a standard Eisenstein series.

Some readers may be (just like ourselves) more acquainted with pseudodifferential analysis than with modular form theory: some very accessible reading about theta series is to be found in [23, Chap. 6] or [15, Chaps. 10–11]; Chap. 8 of [19] is an easy introduction to Poincaré series (for more general subgroups of $SL(2,\mathbb{Z})$ and multipliers).

It is perfectly possible to give a Poincaré-style formula for the distribution $\mathfrak{C}$ itself, not only for its isotypic components: to do this, however, it is necessary, so as to make everything explicit, to characterize this (tempered) distribution in terms of its quadratic $\mathcal{M}$-transform, as introduced in (2.1.3). Since we are dealing with an even distribution, only $(\mathcal{M}h)_0$ is needed, and we now do the calculation to show the advantage of tracing the metaplectic representation by means of the global formula (2.1.4).

To make the computation, we may assume that σ has real entries: let σ_1 and σ_2 be its two (positive) eigenvalues. According to [21, p. 83], one has

$$J_0(2\pi|x|) = \frac{1}{2i\pi} \int_{-\infty}^{(0^+)} e^{\pi(t - \frac{|x|^2}{t})} \frac{dt}{t}, \tag{5.1.26}$$

where the contour is that which starts from $-\infty$ and returns there after having turned around 0 once in the counterclockwise direction. Since

$$\int_{\mathbb{R}^2} e^{-\pi\langle\sigma x, x\rangle} e^{-\frac{\pi|x|^2}{t}} dx - (\sigma_1 + t^{-1})^{-\frac{1}{2}} (\sigma_2 + t^{-1})^{-\frac{1}{2}} \tag{5.1.27}$$

(for large $|t|$, the determination of the square roots is close to real positive numbers), one has

$$(\mathcal{M}h)_0(\sigma) = \frac{1}{2i\pi} \int_{-\infty}^{(0^+)} e^{\pi t} (\sigma_1 + t^{-1})^{-\frac{1}{2}} (\sigma_2 + t^{-1})^{-\frac{1}{2}} \frac{dt}{t} : \tag{5.1.28}$$

changing the contour if needed, it is no loss of generality to assume that $|t| > \frac{\mathrm{tr}\,\sigma}{\det\sigma}$ for t lying on it.

Now, for $x \geq \frac{1}{2}$ and $z \in \mathbb{C}$, $|z| < \frac{1}{2x}$, one has

$$(1 - 2xz + z^2)^{-\frac{1}{2}} = \sum_{j \geq 0} P_j(x) z^j, \tag{5.1.29}$$

where (P_j) is the sequence of Legendre polynomials. Indeed, as $|z| < 1$, this identity holds [21, p. 232] if $\frac{1}{2} \leq x < 1$, while both sides are analytic functions of (x, z) (in the real sense with respect to x) in the domain indicated: the left-hand side because the two zeros of the polynomial $z \mapsto 1 - 2xz + z^2$, to wit $(x + \sqrt{x^2 - 1})^{\pm 1}$ if $x \geq 1$, and complex numbers on the unit circle if $\frac{1}{2} \leq x < 1$, both have absolute values $> \frac{1}{2x}$, the right-hand side in view of the estimate [21, p. 235] $0 < P_j(x) \leq C j^{-\frac{1}{2}} (2x)^j$ if $x > 1$. Hence, for t on the above-specified contour,

$$(\sigma_1 + t^{-1})^{-\frac{1}{2}} (\sigma_2 + t^{-1})^{-\frac{1}{2}} = \sum_{j \geq 0} P_j \left(-\frac{\mathrm{tr}\,\sigma}{2(\det\sigma)^{\frac{1}{2}}} \right) (\det\sigma)^{-\frac{j+1}{2}} t^{-j}, \tag{5.1.30}$$

so that

$$(\mathcal{M}h)_0(\sigma) = \sum_{j \geq 0} \frac{\pi^j}{j!} \, P_j \left(-\frac{\mathrm{tr}\,\sigma}{2\,(\det \sigma)^{\frac{1}{2}}} \right) (\det \sigma)^{-\frac{j+1}{2}}. \tag{5.1.31}$$

With the help of the same reference again, we obtain

$$(\mathcal{M}h)_0(\sigma) = (\sigma_1 \sigma_2)^{-\frac{1}{2}} e^{-\frac{\pi}{2} \frac{\sigma_1 + \sigma_2}{\sigma_1 \sigma_2}} I_0 \left(\frac{\pi\,(\sigma_1 - \sigma_2)}{2\,\sigma_1\,\sigma_2} \right): \tag{5.1.32}$$

we have used, here, the fact that

$$\frac{(\mathrm{tr}\,\sigma)^2}{4\det \sigma} - 1 = \frac{(\sigma_1 - \sigma_2)^2}{4\,\sigma_1\,\sigma_2}. \tag{5.1.33}$$

We now use (2.1.4), for the special case of the matrix which is the image of $g = \left(\begin{smallmatrix} a & b \\ c & d \end{smallmatrix} \right) \in SL(2,\mathbb{R})$ under the embedding of that group into $\mathrm{Sp}(2,\mathbb{R})$: recall from Sect. 2.1 that, in this case, there is no sign ambiguity. We find

$$(\mathcal{M}\,\mathrm{Met}\left(\left(\begin{smallmatrix} a & b \\ c & d \end{smallmatrix} \right)\right) h)_0(\sigma) = [\det(ib\sigma + d)]^{-\frac{1}{2}} (\mathcal{M}u)((a\sigma - ic)(ib\sigma + d)^{-1}). \tag{5.1.34}$$

The ambiguity arising from the square root is only apparent: when σ has real entries, so that σ_1 and σ_2 are real and positive, the square root of the product $(ib\sigma_1 + d)$ $(ib\sigma_2 + d)$ is obtained as the product of $ib\sigma_2 + d$ by the square root of the number $\frac{ib\sigma_1 + d}{ib\sigma_2 + d}$, which always has a positive real part. Setting

$$\tau_1 = \frac{a\sigma_1 - ic}{ib\sigma_1 + d}, \qquad \tau_2 = \frac{a\sigma_2 - ic}{ib\sigma_2 + d}, \tag{5.1.35}$$

one finds that

$$\frac{\tau_1 - \tau_2}{2\,\tau_1\,\tau_2} = \frac{\sigma_1 - \sigma_2}{2\,(a\sigma_1 - ic)(a\sigma_2 - ic)} \tag{5.1.36}$$

and, if $a \neq 0$,

$$\begin{aligned}
\frac{\tau_1 + \tau_2}{2\,\tau_1\,\tau_2} &= \frac{2iab\,\sigma_1\,\sigma_2 + (ad + bc)\,(\sigma_1 + \sigma_2) - 2icd}{2\,(a\sigma_1 - ic)(a\sigma_2 - ic)} \\
&= \frac{ib}{a} + \frac{\sigma_1 + \sigma_2 - 2i\frac{c}{a}}{2\,(a\sigma_1 - ic)(a\sigma_2 - ic)}
\end{aligned} \tag{5.1.37}$$

so that, if $a \neq 0$,

$$\begin{aligned}
(\mathcal{M}\,\mathrm{Met}\left(\left(\begin{smallmatrix} a & b \\ c & d \end{smallmatrix} \right)\right) h)_0(\sigma) &= ((a\,\sigma_1 - ic)(a\,\sigma_2 - ic))^{-\frac{1}{2}} \\
I_0 &\left(\frac{\pi\,(\sigma_1 - \sigma_2)}{2\,(a\sigma_1 - ic)(a\sigma_2 - ic)} \right) \exp \left(-\pi \left(\frac{ib}{a} + \frac{\sigma_1 + \sigma_2 - 2i\frac{c}{a}}{2\,(a\sigma_1 - ic)(a\sigma_2 - ic)} \right) \right);
\end{aligned} \tag{5.1.38}$$

if $a = 0$,

$$(\mathcal{M}\text{Met}\left(\left(\begin{smallmatrix} a & b \\ c & d \end{smallmatrix}\right)\right) h)_0(\sigma) = \frac{i}{c} I_0\left(\frac{\pi(\sigma_1 - \sigma_2)}{2c^2}\right) \exp\left(-\pi\left(\frac{id}{c} + \frac{\sigma_1 + \sigma_2}{2c^2}\right)\right). \tag{5.1.39}$$

Applying (5.1.20), we thus obtain a Poincaré-type series representation of the $\mathcal{M}$-transform of the $\text{Met}^{(2)}(\Gamma_2)$-automorphic distribution $\mathfrak{C}$: note that, contrary to its isotypic components, $\mathfrak{C}$ cannot be characterized by means of a function of one variable only.

The advantage of the constructions which precede, so far, is limited: starting, say, from the measure (5.1.18) and under the assumption that $\alpha_m = 0$ unless $m \geq 2$, the distribution $\mathfrak{C}$ is an automorphic object (a distribution, invariant under $\text{Met}^{(2)}(\Gamma_2)$), the decomposition of which under the rotation group involves infinitely many Poincaré series of a classical style. This point of view, closer from the one usual when dealing with automorphic functions, not only with nonholomorphic modular forms, is certainly helpful when dealing with bilinear operations, such as the Rankin–Cohen brackets, which do not preserve the weight of (holomorphic) modular forms.

We now explore some links between modular form theory and ascending pseudodifferential analysis, starting with a way, based on the calculus, to associate with any holomorphic modular form of weight $m + 1$ a sequence of nonholomorphic modular forms of weights $m + 1 + 2p$, $p \geq 0$. It is better to start from the consideration of a non-automorphic symbol.

Proposition 5.1.2. *Let $h \in S_m(\mathbb{R}^2)$ with $m \geq 1$, and set $\chi = \Theta_m h$. Assuming that k and p are nonnegative integers and setting $j = k + m + 1 + 2p$, one has the identity*

$$(\phi_z^j \mid \text{Op}^{\text{asc}}(h)\, \phi_z^k) = \pi^{\frac{m+1}{2}} F_m^{j,k} \times \mathcal{M}_p(z; h), \tag{5.1.40}$$

where the constant $F_m^{j,k}$ has been introduced in Lemma 3.2.2 and where

$$\mathcal{M}_p(z; h) = \sum_{r=0}^{p} \frac{(2i)^r}{(m+r)!} \binom{p}{r} (\text{Im } z)^{\frac{m+1+2r}{2}} \chi^{(r)}(z). \tag{5.1.41}$$

For every $g = \left(\begin{smallmatrix} a & b \\ c & d \end{smallmatrix}\right) \in SL(2,\mathbb{R})$, one has the covariance relation

$$\mathcal{M}_p\left(\frac{az+b}{cz+d}; h\right) = \left(\frac{cz+d}{|cz+d|}\right)^{m+1+2p} \mathcal{M}_p(z; \text{Met}^{(2)}(g^{-1})h). \tag{5.1.42}$$

Finally, setting (cf. [4, p. 130]) $\Delta_m = -y^2\left(\frac{\partial^2}{\partial x^2} + \frac{\partial^2}{\partial y^2}\right) + imy\frac{\partial}{\partial x}$, one has

$$\left(\Delta_{j-k} - \frac{1}{4}\right)\left(\phi_z^j \mid \text{Op}^{\text{asc}}(h)\, \phi_z^k\right) = \frac{1}{4}\left(\phi_z^j \mid \text{Op}^{\text{asc}}(\mathcal{R}^2 h)\, \phi_z^k\right), \tag{5.1.43}$$

with $\mathcal{R} = \xi\frac{\partial}{\partial x} - x\frac{\partial}{\partial \xi}$, the rotation operator already considered in Theorem 3.1.8.

Proof. Set

$$\mathcal{N}_{\text{asc}}^{j,k}(z;h) = (\phi_z^j \,|\, \text{Op}^{\text{asc}}(h) \,\phi_z^k). \tag{5.1.44}$$

According to (3.2.51), one has

$$\mathcal{N}_{\text{asc}}^{j,k}(z;h) = \sum_{r=0}^{p} (\text{Im } z)^{\frac{m+1+2r}{2}} \, T_{m+1+2r}^{j,k}(z), \tag{5.1.45}$$

where $T_{m+1+2r}^{j,k}$, as given by (3.2.58), reduces to

$$T_{m+1+2r}^{j,k}(z) = \frac{(2i)^r}{(m+r)!} \binom{p}{r} \pi^{\frac{m+1}{2}} \, F_m^{j,k} \, \chi^{(r)}(z) \tag{5.1.46}$$

because h reduces to its isotypic component h_m. Equation (5.1.40) follows. With $g = \left(\begin{smallmatrix} a & b \\ c & d \end{smallmatrix}\right) \in SL(2,\mathbb{R})$, one obtains from the covariance of the ascending pseudodifferential analysis that

$$(\text{Ana}(g)\,\phi_z^j \,|\, \text{Op}^{\text{asc}}(h)\,\text{Ana}(g)\,\phi_z^k) = (\phi_z^j \,|\, \text{Op}^{\text{asc}}\left(\text{Met}^{(2)}(g^{-1})(h)\right)\phi_z^k). \tag{5.1.47}$$

On the other hand, as a consequence of (2.2.31),

$$(\text{Ana}(g)\,\phi_z^j \,|\, \text{Op}^{\text{asc}}(h)\,\text{Ana}(g)\,\phi_z^k) = \left(\frac{cz+d}{|cz+d|}\right)^{k-j} \left(\phi_{\frac{az+b}{cz+d}}^j \,\Big|\, \text{Op}^{\text{asc}}(h)\,\phi_{\frac{az+b}{cz+d}}^k\right)$$

$$= \left(\frac{cz+d}{|cz+d|}\right)^{k-j} \mathcal{N}_{\text{asc}}^{j,k}\left(\frac{az+b}{cz+d};h\right). \tag{5.1.48}$$

Comparing (5.1.56) and (5.1.48), one obtains (5.1.42).

Setting, as in [4, p. 129], $R_k = (z - \bar{z})\frac{\partial}{\partial \bar{z}} + \frac{k}{2}$, one verifies the equation

$$\mathcal{M}_{p+1}(z;h) = \frac{1}{m+1+p} R_{m+1+2p} \, \mathcal{M}_p(z;h): \tag{5.1.49}$$

starting from the equation

$$\Delta_{m+1} \, \mathcal{M}_0(z;h) = \frac{1-m^2}{4} \, \mathcal{M}_0(z;h), \tag{5.1.50}$$

easily checked in a direct way, one obtains the equation

$$\Delta_{m+1+2p} \, \mathcal{M}_p(z;h) = \frac{1-m^2}{4} \, \mathcal{M}_p(z;h), \tag{5.1.51}$$

equivalent to (5.1.43), by induction, using the identity $\Delta_{k+2} R_k = R_k \Delta_k$ indicated in [4, p. 143]. $\qquad\square$

Corollary 5.1.3. *Let χ be a modular form of weight $m+1$ for some arithmetic group $\Gamma \subset SL(2,\mathbb{R})$ and character κ. Set, for $p = 0, 1, \ldots,$*

$$\mathcal{M}_p(z; \chi) = \sum_{r=0}^{p} \frac{(2i)^r}{(m+r)!} \binom{p}{r} (\operatorname{Im} z)^{\frac{m+1+2r}{2}} \chi^{(r)}(z): \qquad (5.1.52)$$

for every $\left(\begin{smallmatrix} a & b \\ c & d \end{smallmatrix}\right) \in \Gamma$, one then has

$$\mathcal{M}_p\left(\frac{az+b}{cz+d}; \chi\right) = \kappa\left(\left(\begin{smallmatrix} a & b \\ c & d \end{smallmatrix}\right)\right) \left(\frac{cz+d}{|cz+d|}\right)^{m+1+2p} \mathcal{M}_p(z; \chi). \qquad (5.1.53)$$

On the other hand,

$$\Delta_{m+1+2p}\, \mathcal{M}_p(z; \chi) = \frac{1 - m^2}{4}\, \mathcal{M}_p(z; \chi): \qquad (5.1.54)$$

in other words, $z \mapsto \mathcal{M}_p(z; \chi)$ is a nonholomorphic modular form of weight $m + 1 + 2p$ in the sense of [4, p. 135].

Proof. From Proposition 2.1.1,

$$\Theta_m \operatorname{Met}^{(2)}(g)(h) = \mathcal{D}_{m+1}(g^{-1})(\Theta_m h), \qquad (5.1.55)$$

so that, if one agrees to denote from now on $\mathcal{N}_{\mathrm{asc}}^{j,k}(z; h)$ as $\mathcal{N}_m^{j,k}(z; \chi)$ (beware. in the new notation, one must keep track of m), (5.1.42) may be rewritten as

$$\left(\frac{cz+d}{|cz+d|}\right)^{k-j} \mathcal{N}_m^{j,k}\left(\frac{az+b}{cz+d}; \chi\right) = \mathcal{N}_m^{j,k}(z; \mathcal{D}_{m+1}(g^{-1})\chi). \qquad (5.1.56)$$

If one makes the transformation $\mathcal{D}_{m+1}(g^{-1})$ explicit, this is a functional identity which continues to be true for arbitrary holomorphic functions χ on Π and for every $g \in SL(2,\mathbb{R})$, $\mathcal{N}_m^{j,k}(z; k)$ being linked to $\mathcal{M}_p(z; \chi)$ by (5.1.40). If χ is a modular form of weight $m+1$ for the group Γ and for some character κ, one has $\mathcal{D}_{m+1}(\gamma^{-1})\chi = \kappa(\gamma)\chi$ for every $\gamma \in \Gamma$, which leads to (5.1.53). $\qquad \square$

Note that the corollary could have been proved just as easily without appealing to the operator calculus interpretation. However, the latter one makes it possible to interpret the Rankin–Cohen brackets of two modular forms as the terms arising from the decomposition into isotypic components of the product of associated operators. Building associative algebras, defined with the help of Rankin–Cohen brackets, of formal series of modular forms, was done in [7]. Independently, in [34], a certain symbolic calculus led to closed algebras of Hilbert–Schmidt operators, the associated composition of symbols expressing itself by means of Rankin–Cohen brackets. Our present point of view is very close to that of the second reference: only, the phase space is now $\mathbb{R}^2$ (instead of a one-sheeted hyperboloid). More important, it has led to the ascending pseudodifferential analysis, the existence of which we had suspected for years, but whose construction would not have been possible before anaplectic analysis (built for other purposes) had been developed.

As a consequence of the composition formula in ascending pseudodifferential analysis, one can establish nontrivial relations between the operators $\chi \mapsto \mathcal{M}_p(.;\chi)$ or, what amounts to the same in view of (5.1.40), $\chi \mapsto \mathcal{N}_m^{j,k}(.;\chi)$, and Rankin–Cohen brackets.

Proposition 5.1.4. *Given two holomorphic functions χ_1 and χ_2 on Π, and two positive integers m_1 and m_2, set $m = m_1 + m_2 + 1$ and*

$$\chi_{m+2\ell+1} = \left(\frac{i}{p}\right)^\ell K_{m+2\ell+1}^{m_1+1,m_2+1}(\chi^1,\chi^2) \tag{5.1.57}$$

for every $\ell = 0,1,\ldots$, where the right-hand side involves the Rankin–Cohen bracket recalled in (3.4.16). With $c_n = \frac{2^{2n} n!}{(2n)!}$, one has, for $p = 0,1,\ldots$ and $k = 0,1,\ldots$, the identity

$$\sum_{\ell=0}^{p} \mathcal{N}_{m+2\ell}^{k+m+1+2p,k}(z;\chi_{m+2\ell+1}) = \sum_{q=0}^{p} c_{k+m_2+1+2q}$$
$$\times \mathcal{N}_{m_1}^{k+m+1+2p,k+m_2+1+2q}(z;\chi^1)\,\mathcal{N}_{m_2}^{k+m_2+1+2q,k}(z;\chi^2). \tag{5.1.58}$$

Proof. To prove this purely formal identity between two bidifferential operators, we may assume that, with the notation of Lemma 3.4.1, one has $\chi^1 = \Theta_{m_1} h^1$ and $\chi^2 = \Theta_{m_2} h^2$ for some $h^1 \in \mathcal{G}_{m_1}(\mathbb{R}^2)$ and $h^2 \in \mathcal{G}_{m_2}(\mathbb{R}^2)$. Set $B_1 = \mathrm{Op}_{m_1}^{\mathrm{asc}}(h^1)$ and $B_2 = \mathrm{Op}_{m_2}^{\mathrm{asc}}(h^2)$. From Theorem 3.4.5, one has $B_1 B_2 = \mathrm{Op}^{\mathrm{asc}}(h)$ with $h = \sum_{\ell \geq 0} h_{m+2\ell}$, where

$$\Theta_{m+2\ell} h_{m+2\ell} = \chi_{m+2\ell+1}. \tag{5.1.59}$$

In view of the definition (5.1.44) of the function $\mathcal{N}_{m+2\ell}^{j,k}(z;\chi_{m+2\ell+1})$, the identity to be proven coincides with the identity (3.4.2) after we have set $j = k+m+1+2p$ and made the following observations regarding the summation indices there: first, that only the terms $h_{m+2\ell}$ with $0 \leq \ell \leq p$ can contribute to the scalar product $(\phi_z^j | B_1 B_2 \phi_z^k)$; next, that $k+m_2+1 \leq n \leq j-m_1-1 = k+m_2+1+2p$ as already noted in the beginning of the proof of Lemma 3.4.1, and that $n-m_2-k-1$ must be even if $(\phi_z^n | B_2 \phi_z^k) \neq 0$. $\qquad\square$

Remark 5.1.1. In Theorem 3.4.5, we have shown that the sharp product, in ascending pseudodifferential analysis, of two isotypic symbols, admits a decomposition as a series of isotypic terms, each of which can be interpreted (after intertwining under the Θ-operator) as a Rankin–Cohen bracket of the two symbols one started from.

There is an absolutely similar situation in "usual" pseudodifferential analysis (i.e., Weyl calculus): starting from two homogeneous symbols, and decomposing their sharp product as an integral of homogeneous terms (for spectral-theoretic reasons, the degrees must lie on the line $-1+i\mathbb{R}$), we obtain a covariant – if only partly defined – way to associate with two distributions on the plane, homogeneous of degrees $-1-i\lambda_1$ and $-1-i\lambda_2$, a family of distributions homogeneous of degrees $-1-i\lambda$. Now, one can transform distributions on $\mathbb{R}^2$ into pairs of functions on the half-plane, under the map $\mathfrak{S} \mapsto (f_0, f_1)$ indicated in the beginning of the present section. All this is still meaningful in the automorphic situation [36]: of course, integral superpositions of Eisenstein distributions (on $\mathbb{R}^2$) or Eisenstein series (on Π) do not suffice any more, and it is then necessary to add series of cusp distributions, or cusp forms. The sharp product, in the automorphic Weyl calculus, of two modular distributions, can be made explicit as a combination of integrals and series, involving the nonholomorphic analogues of Rankin–Cohen brackets.

Some remarks, however, are in order. First, the nonholomorphic theory is considerably more difficult than the holomorphic one since the analogues of Rankin–Cohen brackets do no longer express themselves with the help of bidifferential operators: operators with singular integral kernels are needed, on $\mathbb{R}^2$ as well as on Π [35]. Also, in the automorphic environment, the correct sharp composition of symbols has nothing to do with the one (involving so-called Moyal brackets) universally known: as an example, the sharp composition, in the Weyl calculus, of two Eisenstein distributions can be given a meaning – and can be computed – whereas no term from the usual series, for instance the pointwise product, can make sense. The formula for the sharp composition of symbols of the Weyl calculus relevant to the present analysis, whether in the automorphic case or not, is to be found in [36, Sect. 17]: as indicated above, it is based on the decomposition of symbols into their homogeneous components of degrees lying on the line $-1 + i\mathbb{R}$, not on polynomial approximation or on Taylor's formula. It is not surprising that only this kind of composition formula could be helpful in an arithmetic environment: for instance, it is the only one that can be generalized [3] to p-adic analysis, in which no differential operators can exist.

Within the realm of automorphic distributions for the full modular group, extensive calculations have been made [36] toward the aim of obtaining the spectral decompositions of sharp products of modular distributions (Eisenstein distributions and cusp distributions) in a fully explicit way. To a newcomer to modular form theory (like the present author), these calculations had the advantage of bringing a large amount of number-theoretic concepts such as the Hecke operators, L-functions and convolution L-functions, Eulerian products, etc., into the picture. We hope that, eventually, something analogous can be obtained from the present theory. Even more so, some important tools, such as Rankin–Selberg integrals [15, p. 232], are based on the simultaneous use of modular forms of the holomorphic and nonholomorphic species: it would be nice to know if this, again, has an interpretation in terms of some appropriate pseudodifferential analysis.

5.2 Moyal Brackets and Rankin–Cohen Brackets

As is well known [27], holomorphic modular forms for the full modular group can also be viewed as homogeneous functions of lattices, holomorphic in a certain sense: this point of view is especially helpful in the description of Hecke operators. Under this correspondence, Rankin–Cohen brackets transfer to Moyal brackets. This is immediate: our point is to see why this does not admit a bona fide interpretation in terms of pseudodifferential analysis; more precisely, that it does so (cf. very end of this section) only in connection with a less satisfactory version of the alternative pseudodifferential analysis, based on the use of a nonlinear phase space. A detour through the Weyl pseudodifferential calculus, in which Moyal brackets are of course well known to be relevant, will help.

For the sake of comparison, we first indicate the analogue, in usual pseudodifferential analysis, of Proposition 5.1.2. Just as in Theorem 3.1.8, it will be seen that, again, it is only a matter of trading the operators $\mathcal{R}$ and $2\pi\mathcal{E} = -i\left(x\frac{\partial}{\partial x} + \xi\frac{\partial}{\partial \xi} + 1\right)$ with each other.

We first need to introduce the L^2-normalized ground state $u(t) = 2^{\frac{1}{4}} e^{-\pi t^2}$ of the standard harmonic oscillator L and, with the help of the creation operator $A^* = \pi^{\frac{1}{2}}(Q - iP)$, the complete orthogonal set of normalized eigenstates u^j, $j \geq 0$, with $u^j = (j!)^{-\frac{1}{2}}(A^*)^j u$. Then, if $z = x + iy \in \Pi$, we set

$$u_z^j(t) = \left(\frac{-iz}{|z|}\right)^{j+\frac{1}{2}} \frac{y^{\frac{1}{4}}}{|z|^{\frac{1}{2}}} u^j\left(\frac{y^{\frac{1}{2}} t}{|z|}\right) e^{i\pi \frac{x}{|z|^2} t^2} : \tag{5.2.1}$$

Proposition 4.2.3 and (4.1.31) show that $u_z^j = \mathrm{Met}^{(1)}(g_z) u^j$ for some element g_z of the metaplectic group above the matrix $\begin{pmatrix} y^{\frac{1}{2}} & y^{-\frac{1}{2}}x \\ 0 & y^{-\frac{1}{2}} \end{pmatrix}$, to wit the one defined right after (4.1.28). To fully realize the close relationship between (5.1.42) and (5.2.3) below, recall that $\mathfrak{S} \circ g = \pi(g^{-1})\mathfrak{S}$ if π is the quasiregular representation of $SL(2,\mathbb{R})$ in $\mathcal{S}'(\mathbb{R}^2)$.

Proposition 5.2.1. *Given* $g = \begin{pmatrix} a & b \\ c & d \end{pmatrix}$, $\mathfrak{S} \in \mathcal{S}'(\mathbb{R}^2)$, *and* $j, k = 0, 1, \ldots$, *the function*

$$\mathcal{N}^{j,k}(z; \mathfrak{S}) = (u_z^j \mid \mathrm{Op}(\mathfrak{S}) u_z^k) \tag{5.2.2}$$

satisfies the covariance equation

$$\mathcal{N}^{j,k}\left(\frac{az+b}{cz+d}; \mathfrak{S}\right) = \left(\frac{cz+d}{|cz+d|}\right)^{j-k} \mathcal{N}^{j,k}(z; \mathfrak{S} \circ g). \tag{5.2.3}$$

Moreover,

$$\left(\Delta_{j-k} - \frac{1}{4}\right)\left(u_z^j \mid \mathrm{Op}(\mathfrak{S}) u_z^k\right) = \left(u_z^j \mid \mathrm{Op}(\pi^2 \mathcal{E}^2 \mathfrak{S}) u_z^k\right). \tag{5.2.4}$$

Proof. Using the covariance of the Weyl calculus under the metaplectic representation, one finds

$$\mathcal{N}^{j,k}(z; \mathfrak{S} \circ g) = \left(u_z^j \mid \mathrm{Met}^{(1)}(\tilde{g}^{-1}) \mathrm{Op}(\mathfrak{S}) \mathrm{Met}^{(1)}(\tilde{g}) u_z^k\right)$$

$$\left(\mathrm{Met}^{(1)}(\tilde{g}) u_z^j \mid \mathrm{Op}(\mathfrak{S}) \mathrm{Met}^{(1)}(\tilde{g}) u_z^k\right) \tag{5.2.5}$$

for any point $\tilde{g}$ of the metaplectic group lying above g. On the other hand, the element $\tilde{g} g_z$ of the metaplectic group lies above the matrix

$$\begin{pmatrix} a & b \\ c & d \end{pmatrix} \begin{pmatrix} y^{\frac{1}{2}} & y^{-\frac{1}{2}} x \\ 0 & y^{-\frac{1}{2}} \end{pmatrix} = \begin{pmatrix} y'^{\frac{1}{2}} & y'^{-\frac{1}{2}} x' \\ 0 & y'^{-\frac{1}{2}} \end{pmatrix} \begin{pmatrix} \frac{cx+d}{|cz+d|} & -\frac{cy}{|cz+d|} \\ \frac{cy}{|cz+d|} & \frac{cx+d}{|cz+d|} \end{pmatrix}, \tag{5.2.6}$$

where we have set $z' = x' + iy' = \frac{az+b}{cz+d}$ on the right-hand side: the verification is straightforward. Now, the first matrix on the right-hand side is covered by the element $g_{z'}$ of the metaplectic group, and the second one can be written as $\begin{pmatrix} \cos\theta & \sin\theta \\ -\sin\theta & \cos\theta \end{pmatrix}$ if one sets $e^{-i\theta} = \frac{cz+d}{|cz+d|}$: covering this matrix by the element $\exp\theta \begin{pmatrix} 0 & 1 \\ -1 & 0 \end{pmatrix}$ of the metaplectic group, and using the case $\nu = 0$ of (4.1.30), one obtains, for an appropriate choice of $\tilde{g}$ and of the square root on the right-hand side,

$$\begin{aligned} \mathrm{Met}^{(1)}(\tilde{g} g_z) u^j &= \left(\frac{cz+d}{|cz+d|} \right)^{j+\frac{1}{2}} \mathrm{Met}^{(1)}(g_{z'}) u^j \\ &= \left(\frac{cz+d}{|cz+d|} \right)^{j+\frac{1}{2}} u^j_{z'}. \end{aligned} \tag{5.2.7}$$

$\square$

Using the same calculation, starting with u^k instead, one obtains (5.2.3): the ambiguity, by the factor ± 1, due to the square root, disappears in the scalar product.

The proof of (5.2.4) begins with the use of (3.1.33), written as

$$A\,\mathrm{Op}(\mathfrak{S})A^* - A^*\,\mathrm{Op}(\mathfrak{S})A = \mathrm{Op}(2i\pi\,\mathcal{E}\,\mathfrak{S}). \tag{5.2.8}$$

One has

$$(u^j \,|\, \mathrm{Op}(2i\pi\,\mathcal{E}\,\mathfrak{S}) u^k) = (A^* u^j \,|\, \mathrm{Op}(\mathfrak{S}) A^* u_k) - (A u^j \,|\, \mathrm{Op}(\mathfrak{S}) A u_k)$$
$$= ((j+1)(k+1))^{\frac{1}{2}} (u^{j+1} \,|\, \mathrm{Op}(\mathfrak{S}) u^{k+1}) - (jk)^{\frac{1}{2}} (u^{j-1} \,|\, \mathrm{Op}(\mathfrak{S}) u^{k-1}). \tag{5.2.9}$$

Since $u^j_z = \mathrm{Met}^{(1)}(g_z) u^j$, the covariance of the Weyl calculus, together with the fact that the Euler operator commutes with the linear action of $SL(2, \mathbb{R})$ on the plane, shows that

$$\begin{aligned} (u^j_z \,|\, \mathrm{Op}(2i\pi\,\mathcal{E}\,\mathfrak{S}) u^k_z) \\ = ((j+1)(k+1))^{\frac{1}{2}} (u^{j+1}_z \,|\, \mathrm{Op}(\mathfrak{S}) u^{k+1}_z) - (jk)^{\frac{1}{2}} (u^{j-1}_z \,|\, \mathrm{Op}(\mathfrak{S}) u^{k-1}_z) \end{aligned} \tag{5.2.10}$$

for every $z \in \Pi$. As a consequence,

$$\begin{aligned} (u^j_z \,|\, \mathrm{Op}(\pi^2\,\mathcal{E}^2\,\mathfrak{S}) u^k_z) &= -\frac{1}{4} ((j+1)(k+1)(j+2)(k+2))^{\frac{1}{2}} (u^{j+2}_z \,|\, \mathrm{Op}(\mathfrak{S}) u^{k+2}_z) \\ &+ \frac{2jk+j+k+1}{4} (u^j_z \,|\, \mathrm{Op}(\mathfrak{S}) u^k_z) - \frac{1}{4} (jk(j-1)(k-1))^{\frac{1}{2}} (u^{j-2}_z \,|\, \mathrm{Op}(\mathfrak{S}) u^{k-2}_z). \end{aligned} \tag{5.2.11}$$

To compute the left-hand side of the claimed identity (5.2.4), we need a lemma.

Lemma 5.2.2. *With $z = x + iy$ and*

$$R_m = 2iy\frac{\partial}{\partial z} + \frac{m}{2}, \qquad L_m = -2iy\frac{\partial}{\partial \bar{z}} - \frac{m}{2}, \qquad (5.2.12)$$

as defined in [4, p. 129], one has

$$R_0 u_z^k = -\frac{(k(k-1))^{\frac{1}{2}}}{2} u_z^{k-2} + (\frac{k}{2} + \frac{1}{4}) u_z^k,$$

$$L_0 u_z^k = \frac{((k+1)(k+2))^{\frac{1}{2}}}{2} u_z^{k+2} - (\frac{k}{2} + \frac{1}{4}) u_z^k. \qquad (5.2.13)$$

Proof. The first two even normalized eigenfunctions of the standard harmonic oscillator are

$$u^0(t) = 2^{\frac{1}{4}} e^{-\pi t^2}, \qquad u^2(t) = 2^{-\frac{1}{2}} (4\pi t^2 - 1) u^0(t). \qquad (5.2.14)$$

Applying (5.2.1), one obtains

$$u_z^0(t) = (2y)^{\frac{1}{4}} (i\bar{z})^{-\frac{1}{2}} e^{\frac{i\pi t^2}{\bar{z}}}, \qquad u_z^2(t) = u_z^0(t) \times 2^{-\frac{1}{2}} \left[\frac{z}{\bar{z}} - 4\pi \frac{yt^2}{\bar{z}^2}\right]; \qquad (5.2.15)$$

it follows, after a straightforward computation, that

$$(R_0 - \frac{1}{4}) u_z^0 = 0, \qquad (L_0 + \frac{1}{4}) u_z^0 = 2^{-\frac{1}{2}} u_z^2. \qquad (5.2.16)$$

Next, from (4.1.26) (recall from Proposition 4.2.3 that the metaplectic representation, when applied to Hermite-like functions, coincides with a restriction of the 0-anaplectic representation), one has

$$A_z^* u_z^k = (k+1)^{\frac{1}{2}} y^{\frac{1}{2}} u_z^{k+1}, \qquad A_z u_z^k = k^{\frac{1}{2}} y^{\frac{1}{2}} u_z^{k-1}. \qquad (5.2.17)$$

When the operators which follow are applied to functions of (t, z), one has the following commutation rules:

$$[L_0, y^{-\frac{1}{2}} A_z^*] = [-2iy\frac{\partial}{\partial \bar{z}}, y^{-\frac{1}{2}}] A_z^* = -\frac{1}{2} y^{-\frac{1}{2}} A_z^*,$$

$$[R_0, y^{-\frac{1}{2}} A_z^*] = [2iy\frac{\partial}{\partial z}, y^{-\frac{1}{2}}] A_z^* + y^{-\frac{1}{2}} (A_z^* - A_z) = y^{-\frac{1}{2}} (\frac{1}{2} A_z^* - A_z). \qquad (5.2.18)$$

Equations (5.2.13) follow by induction, starting from (5.2.16) and using (5.2.17) as well as (5.2.18). □

End of Proof of Proposition 5.2.1. One has

$$L_0 (u_z^j | \mathrm{Op}(\mathfrak{S}) u_z^k) = (R_0 u_z^j | \mathrm{Op}(\mathfrak{S}) u_z^k) + (u_z^j | \mathrm{Op}(\mathfrak{S}) L_0 u_z^k). \qquad (5.2.19)$$

Making the right-hand side explicit with the help of Lemma 5.2.2 and subtracting $\frac{j-k}{2}$ from L_0, one obtains

$$L_{j-k}\left(u_z^j \,|\, \text{Op}(\mathfrak{S})\, u_z^k\right)$$
$$= \frac{1}{2}\left[-(j(j-1))^{\frac{1}{2}}\left(u_z^{j-2}\,|\,\text{Op}(\mathfrak{S})\,u_z^k\right)+((k+1)(k+2))^{\frac{1}{2}}\left(u_z^j\,|\,\text{Op}(\mathfrak{S})\,u_z^{k+2}\right)\right].$$
$$(5.2.20)$$

In a similar way,

$$R_{j-k}\left(u_z^j\,|\,\text{Op}(\mathfrak{S})\,u_z^k\right)$$
$$= \frac{1}{2}\left[((j+1)(j+2))^{\frac{1}{2}}\left(u_z^{j+2}\,|\,\text{Op}(\mathfrak{S})\,u_z^k\right)-(k(k-1))^{\frac{1}{2}}\left(u_z^j\,|\,\text{Op}(\mathfrak{S})\,u_z^{k-2}\right)\right].$$
$$(5.2.21)$$

Using the equation [4, p. 130]

$$\Delta_{j-k}-\frac{1}{4}=-R_{j-k-2}L_{j-k}-\frac{(j-k-1)^2}{4},\qquad(5.2.22)$$

and expanding, one obtains the right-hand side of the claimed identity, as computed in (5.2.11).   □

Remark 5.2.1. Despite their similarity, Propositions 5.1.2 and 5.2.1 show profound differences too. On one hand, the differential equation (5.1.43) (resp., (5.2.4)) relates to the decomposition of a distribution on $\mathbb{R}^2$ into its isotypic (resp., homogeneous) components; on the other hand, a symbol $\mathfrak{S}$ from the usual pseudodifferential calculus is perfectly determined from the knowledge of only four of the functions $\mathcal{N}^{j,k}(.\,;\mathfrak{S})$ (one pair (j,k) with $j\equiv\varepsilon_1$ mod 2 and $k\equiv\varepsilon_2$ is needed for each pair $(\varepsilon_1,\varepsilon_2)\in\{0,1\}\times\{0,1\}$), while infinitely many functions $\mathcal{N}^{j,k}_{\text{asc}}(.\,;h)$ are needed to determine h, as only those with $j-k\geq m+1$ can contribute to the knowledge of the mth isotypic component of h. It has been observed in [4, p. 145] that Maass forms with possibly nonzero weight are of two species: some correspond to holomorphic modular forms and others can be reduced to the weight 0 or 1. Propositions 5.1.2 and 5.2.1 relate the two species to two different symbolic calculi.

In Proposition 5.2.3, one should observe the fundamental difference between the Rankin–Cohen brackets and the *mock* Rankin–Cohen brackets below: changing $m+1$ to $-m$ is not a benign operation. It is impossible to relate the *true* Rankin–Cohen brackets to the Weyl calculus: they are related to the alternative symbolic calculus.

Proposition 5.2.3. Fix $m=0,1,\ldots$, and let $\mathfrak{S}\in\text{Pol}(m)$, the linear space of polynomials on $\mathbb{R}^2$ homogeneous of degree m. Choosing an arbitrary $j=0,1,\ldots$, set

$$(\theta\mathfrak{S})(z)=(2i)^m\,\pi^{\frac{m}{2}}\left[\frac{j!}{(j+m)!}\right]^{\frac{1}{2}}(\text{Im}\,z)^{\frac{m}{2}}\left(u_z^j\,|\,\text{Op}(\mathfrak{S})\,u_z^{j+m}\right).\qquad(5.2.23)$$

One has simply

$$(\theta\mathfrak{S})(z) = \mathfrak{S}(z,1), \qquad z \in \Pi. \tag{5.2.24}$$

On the other hand, given $\mathfrak{S}^1 \in \mathrm{Pol}(m_1)$ and $\mathfrak{S}^2 \in \mathrm{Pol}(m_2)$, the component $\mathrm{Pol}(m_1 + m_2 - 2p)$ of the expansion of the sharp product $\mathfrak{S}^1 \# \mathfrak{S}^2$ as a sum of homogeneous polynomials, to wit

$$\mathfrak{S}_{m_1+m_2-2p}(x,\xi) = \left(\frac{1}{4i\pi}\right)^p \sum_{q=0}^{p} \frac{(-1)^q}{q!(p-q)!} \left[\left(\frac{\partial}{\partial x}\right)^q \left(\frac{\partial}{\partial \xi}\right)^{p-q} \mathfrak{S}^1\right](x,\xi)$$

$$\left[\left(\frac{\partial}{\partial x}\right)^{p-q} \left(\frac{\partial}{\partial \xi}\right)^{q} \mathfrak{S}^2\right](x,\xi), \tag{5.2.25}$$

transfers through θ to $\left(\frac{i}{4\pi}\right)^p$ times the mock Rankin–Cohen bracket defined, if $\chi^1 = \theta\mathfrak{S}^1$, $\chi^2 = \theta\mathfrak{S}^2$, as

$$\left(\widetilde{\mathcal{K}}^{m_1,m_2}_{m_1+m_2-2p}(\chi^1,\chi^2)\right)(z) = \sum_{q=0}^{p} (-1)^q \binom{m_1-p+q}{q}\binom{m_2-q}{p-q} (\chi^1)^{(p-q)}(z)(\chi^2)^{(q)}(z). \tag{5.2.26}$$

Proof. As an immediate consequence of (5.2.3), one has

$$(\theta\mathfrak{S})\left(\frac{az+b}{cz+d}\right) = (cz+d)^{-m}\left[\theta(\mathfrak{S} \circ g)\right](z) \tag{5.2.27}$$

for every $g = \left(\begin{smallmatrix} a & b \\ c & d \end{smallmatrix}\right)$. For comparison, recall that (2.1.11) reads

$$(\Theta_m h)\left(\frac{az+b}{cz+d}\right) = (cz+d)^{m+1}\left[\Theta_m(\mathrm{Met}^{(2)}(g^{-1})h)\right](z), \tag{5.2.28}$$

an identity with a quite different exponent. We now prove (5.2.24). Given $\mathfrak{S} \in \mathrm{Pol}(m)$, set, with $\Delta = -\left(\frac{\partial^2}{\partial x^2} + \frac{\partial^2}{\partial \xi^2}\right)$,

$$\mathfrak{T} = \left(\exp\left(-\frac{\Delta}{8\pi}\right)\right)\mathfrak{S} = \sum_{\ell \geq 0} \frac{1}{\ell!}\left(-\frac{\Delta}{8\pi}\right)^\ell \mathfrak{S}: \tag{5.2.29}$$

this is of course a nonhomogeneous polynomial of degree m, called the Wick symbol of the operator $\mathrm{Op}(\mathfrak{S})$, as may be found in many places [33], though possibly not with the same normalization: this means that, if

$$\mathfrak{T}(x,\xi) = \sum_{\alpha,\beta} a_{\alpha,\beta}(x+i\xi)^\alpha(x-i\xi)^\beta, \tag{5.2.30}$$

one has

$$\mathrm{Op}(\mathfrak{S}) = \sum_{\alpha,\beta} \pi^{-\frac{\alpha+\beta}{2}} a_{\alpha,\beta} A^{*\beta} A^\alpha. \tag{5.2.31}$$

Hence,

$$
(u^j \,|\, \mathrm{Op}(\mathfrak{S})\, u^{j+m}) = \sum_{\alpha,\beta} \pi^{-\frac{\alpha+\beta}{2}}\, a_{\alpha,\beta}\, (A^\beta u^j \,|\, A^\alpha u^{j+m})
$$

$$
= \sum_{\alpha,\beta} \pi^{-\frac{\alpha+\beta}{2}}\, a_{\alpha,\beta}\, \left[\frac{j!\,(j+m)!}{(j-\beta)!\,(j+m-\alpha)!} \right]^{\frac{1}{2}} (u^{j-\beta} \,|\, u^{j+m-\alpha}):
$$

$$(5.2.32)$$

now one has $\alpha \geq 0$, $\beta \geq 0$, $\alpha + \beta \leq m$, and the scalar product on the right-hand side vanishes unless $j - \beta = j + m - \alpha$, i.e., $\alpha = m$, $\beta = 0$, in which case its value is 1. Hence

$$
(u^j \,|\, \mathrm{Op}(\mathfrak{S})\, u^{j+m}) = \pi^{-\frac{m}{2}}\, a_{m,0}\, \left[\frac{(j+m)!}{j!} \right]^{\frac{1}{2}}
$$

$$
= \frac{\pi^{-\frac{m}{2}}}{m!} \left[\frac{(j+m)!}{j!} \right]^{\frac{1}{2}} \left(\frac{1}{2}\left(\frac{\partial}{\partial x} - i \frac{\partial}{\partial \xi} \right) \right)^m \Bigg|_{(x,\xi)=(0,0)} \mathfrak{T}(x,\xi),
$$

$$(5.2.33)$$

and it does not change anything if $\mathfrak{T}$ is replaced by $\mathfrak{S}$ on the right-hand side. In other words,

$$
(\theta\mathfrak{S})(i) = \frac{(2i)^m}{m!} \left(\frac{1}{2}\left(\frac{\partial}{\partial x} - i \frac{\partial}{\partial \xi} \right) \right)^m \Bigg|_{(x,\xi)=(0,0)} \mathfrak{S}(x,\xi)
$$

$$
= i^m \sum_{r=0}^{m} \frac{1}{r!\,(m-r)!} \left(\left(\frac{\partial}{\partial x} \right)^r \left(\frac{\partial}{\partial \xi} \right)^{m-r} \mathfrak{S} \right)(0,0) \cdot (-i)^{m-r}
$$

$$
= \mathfrak{S}(i,1).
$$

$$(5.2.34)$$

This is just the case $z = i$ of (5.2.24). The general case is obtained with the help of (5.2.27).

Let us write the Moyal bracket under consideration as

$$
\mathfrak{S}_{m_1+m_2-2p}(x,\xi) = \frac{(4i\pi)^{-p}}{p!} \left[-\frac{\partial^2}{\partial x_1 \partial \xi_2} + \frac{\partial^2}{\partial x_2 \partial \xi_1} \right]^p \left(\mathfrak{S}^1(x_1,\xi_1)\, \mathfrak{S}^2(x_2,\xi_2) \right),
$$

$$(5.2.35)$$

it being understood that the result has to be evaluated at $(x_1,\xi_1) = (x_2,\xi_2) = (x,\xi)$. Set

$$
x_1 = t_1 z_1,\; \xi_1 = t_1;\quad x_2 = t_2 z_2,\; \xi_2 = t_2.
$$

$$(5.2.36)$$

The operator within brackets on the right-hand side of the last equation has to be replaced by

$$
D = \frac{z_2 - z_1}{t_1 t_2} \frac{\partial^2}{\partial z_1 \partial z_2} - \frac{1}{t_1} \frac{\partial^2}{\partial z_1 \partial t_2} + \frac{1}{t_2} \frac{\partial^2}{\partial z_2 \partial t_1},
$$

$$(5.2.37)$$

and the function $\chi_{m_1+m_2-2p} = \theta\mathfrak{S}_{m_1+m_2-2p}$ can be written as

$$\chi_{m_1+m_2-2p}(z) = \left[D\left(t_1^{m_1}\, t_2^{m_2}\, \chi^1(z_1)\, \chi^2(z_2)\right)\right]^p \quad (t_1=t_2=1,\, z_1=z_2=z)$$

$$= \left[(z_2-z_1)\frac{\partial^2}{\partial z_1\,\partial z_2} - m_2\frac{\partial}{\partial z_1} + m_1\frac{\partial}{\partial z_2}\right]^p \left(\chi^1(z_1)\,\chi^2(z_2)\right)(z_1=z_2=z).$$

$$(5.2.38)$$

The last differential operator under consideration is a sum of partial differential operators the coefficients of which coincide with powers of $z_2 - z_1$: at the point $z_1 = z_2 = z$, this reduces to an operator with constant coefficients. Taking the degrees of the operators into consideration, one obtains, just as in (3.4.18),

$$\chi_{m_1+m_2-2p}(z) = \sum_{q=0}^{p} \gamma_q\, (\chi^1)^{(p-q)}(z)\,(\chi^2)^{(q)}(z). \qquad (5.2.39)$$

To make the expression on the right-hand side explicit, we proceed exactly as in the proof of Lemma 3.4.2, only taking this time $\chi^1(z) = z^{m_1}$, $\chi^2(z) = 1$, which correspond to $\mathfrak{S}^1(x,\xi) = x^{m_1}$ $\mathfrak{S}^2(x,\xi) = \xi^{m_2}$: then, the covariance (5.2.27) of the Weyl calculus (or of the Moyal-type operation) yields that $\chi_{m_1+m_2-2p}(z)$ coincides, up to some constant, with the expression $\widetilde{\mathcal{K}}^{m_1,m_2}_{m_1+m_2-2p}(z)$ as defined in (5.2.26). It is easy to determine the constant of proportionality since, on one hand, it is immediate in the case under study that

$$\left(\widetilde{\mathcal{K}}^{m_1,m_2}_{m_1+m_2-2p}(\chi^1,\chi^2)\right)(z) = \frac{m_2!}{p!\,(m_2-p)!}\frac{m_1!}{(m_1-p)!}\,z^{m_1-p}; \qquad (5.2.40)$$

on the other hand, the definition (5.2.25) of Moyal brackets reduces in our case to

$$\mathfrak{S}_{m_1+m_2+2p}(x,\xi) = \frac{1}{p!}\left(-\frac{1}{4i\pi}\right)^p \left(\frac{\partial}{\partial x}\right)^p \mathfrak{S}^1(x,\xi)\left(\frac{\partial}{\partial \xi}\right)^p \mathfrak{S}^2(x,\xi),$$

$$(5.2.41)$$

which leads to

$$\chi_{m_1+m_2-2p}(z) = \frac{1}{p!}\left(-\frac{1}{4i\pi}\right)^p \frac{m_1!\,m_2!}{(m!-p)!\,(m_2-p)!}\,z^{m_1-p}. \qquad (5.2.42)$$

This concludes the proof of Proposition 5.2.3.   □

Remark 5.2.2. Let us remark that the case of homogeneous polynomials, though a special one, is actually fundamental in the usual pseudodifferential calculus. Most pseudodifferential operator methods are based – in a conscious way or not – on this case, accompanied by localization techniques (for instance, Taylor expansions and estimates). This is the right way to use pseudodifferential analysis in partial differential equations. On the other hand, the different composition formula, in the Weyl calculus, which has been alluded to in Remark 5.1.1, is based on a totally different "harmonic analysis" of functions in the plane, to wit on their decomposition as integrals of families of homogeneous functions. The latter composition formula does not play any role in partial differential equations; however, it is the only one that can be of help when dealing with very singular symbols such as those which occur in automorphic pseudodifferential analysis.

The following easy proposition, unrelated to pseudodifferential analysis, links Rankin–Cohen brackets on the upper half-plane to Moyal brackets not on flat space, but on some domain of $\mathbb{C}^2$.

Proposition 5.2.4. *Let* $p = 0, 1, \ldots$, *and let* Ω *be the domain of* $\mathbb{C}^2$ *consisting of points* (x, ξ) *such that* Im $\frac{x}{\xi} > 0$. *If one identifies, in the usual way [27, p. 135], holomorphic modular forms of weight* $m + 1$ *for the full modular group with holomorphic functions of lattices, homogeneous of degree* $-m - 1$, *in other words* $SL(2, \mathbb{Z})$-*invariant homogeneous holomorphic functions in* Ω, *the (canonical) Rankin–Cohen bracket* $K_{m_1+m_2+2+2p}^{m_1+1,m_2+1}$ *transfers to* $(4i\pi)^p$ *times the Moyal bracket* $(\mathfrak{S}^1, \mathfrak{S}^2) \mapsto \mathfrak{S}_{m_1+m_2+2+2p}$, *formally defined as in* (5.2.35).

Proof. The result is a special case of the following statement, the proof of which is fully identical to the end of the proof of Proposition 5.2.3. The only difference is that the first function from the pair $\chi^1(z) = z^{m_1}$, $\chi^2(z) = 1$ must be replaced by $\chi^1(z) = z^{-m_1-1}$: let $\mathfrak{S}^1$ and $\mathfrak{S}^2$ be holomorphic functions in Ω, respectively, homogeneous of degrees $-m_1 - 1$ and $-m_2 - 1$, where m_1 and m_2 are nonnegative integers. For $p = 0, 1, \ldots$, define the function $\mathfrak{S}_{m_1+m_2+2+2p}$ by the Moyal bracket formula. Defining, in the upper half-plane, the functions

$$\chi^1(z) = \mathfrak{S}^1(z, 1), \quad \chi^2(z) = \mathfrak{S}^2(z, 1), \quad \chi_{m_1+m_2+2+2p}(z) = \mathfrak{S}_{m_1+m_2+2+2p}(z, 1),$$
$$(5.2.43)$$

one has

$$\chi_{m_1+m_2+2+2p} = \left(\frac{1}{4i\pi}\right)^p K_{m_1+m_2+2+2p}^{m_1+1,m_2+1}(\chi^1, \chi^2), \qquad (5.2.44)$$

where the Rankin–Cohen bracket recalled in (3.4.16) occurs on the right-hand side.   □

Remark 5.2.3. It would of course be possible to regard the complex domain Ω as the phase space for the following quantization rule: given a holomorphic function F on this domain, with $F = \sum_{m \geq 1} F_m$, F_m homogeneous of degree $-m - 1$, associate with F the operator $\text{Op}^{\text{asc}}(\sum h_m)$ with $(\Theta_m h_m)(z) = F_m(z)$. Then, the Moyal brackets on Ω would again have an interpretation in terms of the symbolic calculus, as a consequence of Proposition 5.2.4 and Theorem 3.4.5. However, we would lose any simple covariance rule under the Heisenberg representation; besides, the parallel between the ascending symbolic calculus and the Weyl calculus, the existence of which has been one of the major aims of the present work, would be lost as well in the process.

References

1. F. Bayen, M. Flato, C. Fronsdal, A. Lichnerowicz, D. Sternheimer, *Deformation theory and quantization*, Ann. Phys. **11**(1) (1978) 61–151
2. R. Beals, *Characterization of pseudodifferential operators and applications*, Duke Math. J. **44**(1) (1977) 45–57; ibidem, **46**(1) (1979) 215
3. A. Bechata, *Calcul pseudodifférentiel p-adique*, Ann. Fac. Sci. Toulouse Math. Ser. (6) **13**(2) (2004) 179–240
4. D. Bump, *Automorphic Forms and Representations*, Cambridge Series in Advanced Mathematics **55**, Cambridge University Press, Cambridge, 1996
5. P. Cartier, *Quantum mechanical commutation relations and theta functions*, Proc. Symp. Pure Math. **9** (1965) 361–383
6. H. Cohen, *Sums involving the values at negative integers of L-functions of quadratic characters*, Math. Ann. **217** (1975) 271–295
7. P.B. Cohen, Y. Manin, D. Zagier, *Automorphic Pseudodifferential Operators*, in *Algebraic Aspects of Integrable Systems*, Progress in Nonlinear Differential Equations and Applications **26**, Birkhäuser, Boston, 1996, 17–47
8. G. van Dijk, M. Pevzner, *Ring structures for holomorphic discrete series and Rankin–Cohen brackets*, J. Lie Theory **17**(2) (2007) 283–305
9. S. Gelbart, *Examples of dual reductive pairs*, Proc. Symp. Pure Math. **33** (1977) 287–296
10. I.M. Gelfand, M.I. Graev, N.Ya. Vilenkin, *Generalized Functions*, vol. 5, Academic, London, 1966
11. I.S. Gradstein, I.M. Ryshik, *Tables of Series, Products and Integrals*, vol. 2, Verlag Harri Deutsch, Thun-Frankfurt/M, 1981
12. D.A. Hejhal, *The Selberg trace formula and the Riemann zeta function*, Duke Math. J. **43**(3) (1976) 441–482
13. R. Howe, *θ-series and invariant theory*, Proc. Symp. Pure Math. **33** (1977) 275–285
14. R. Howe, *Quantum mechanics and partial differential equations*, J. Funct. Anal. **38**(2) (1980) 188–254
15. H. Iwaniec, *Topics in Classical Automorphic Forms*, Graduate Studies in Mathematics **17**, American Mathematical Society, Providence, 1997
16. A.W. Knapp, *Representation Theory of Semi-Simple Groups*, Princeton University Press, Princeton, 1986
17. M. Kontsevich, *Deformation quantization of Poisson manifolds*, Lett. Math. Phys. **66**(3) (2003) 157–216
18. P.D. Lax, R.S. Phillips, *Scattering Theory for Automorphic Functions*, Annals of Mathematics Studies **87**, Princeton University Press, Princeton, 1976
19. J. Lehner, *Discontinuous Groups and Automorphic Functions*, Mathematical Surveys **8**, American Mathematical Society, Providence, 1964

20. G. Lion, M. Vergne, *The Weil Representation, Maslov Index and Theta Series*, Progress in Mathematics, Birkhäuser, Basel, 1980

21. W. Magnus, F. Oberhettinger, R.P. Soni, *Formulas and Theorems for the Special Functions of Mathematical Physics*, 3rd edition, Springer, Berlin Heidelberg New York, 1966

22. V.F. Molchanov, *Quantization on the imaginary Lobachevskii plane*, Funksional'nyi Analiz, Ego Prilozheniya **14** (1980) 73–74

23. A. Ogg, *Modular Forms and Dirichlet Series*, Benjamin, New York, 1969

24. L. Pukanszky, *The Plancherel formula for the universal covering group of* $SL(\mathbb{R}, 2)$, Math. Ann. **156** (1964) 96–143

25. S. Rallis, G. Schiffmann, *Automorphic forms constructed from the Weil representation: holomorphic case*, Am. J. Math. **100**(2) (1978) 1049–1122

26. R.A. Rankin, *The construction of automorphic forms from the derivatives of a given form*, J. Indian Math. Soc. **20** (1956) 103–116

27. J.P. Serre, *Cours d'Arithmétique*, Presses Universitaires de France, Paris, 1970

28. D. Shale, *Linear symmetries of free boson fields*, Trans. Am. Math. Soc. **103** (1962) 149–167

29. E.M. Stein, G. Weiss, *Introduction to Fourier Analysis in Euclidean Spaces*, Princeton University Press, Princeton, 1971

30. R.S. Strichartz, *Harmonic analysis on hyperboloids*, J. Funct. Anal. **12** (1973) 341–383

31. A. Unterberger, *Oscillateur harmonique et opérateurs pseudodifférentiels*, Ann. Inst. Fourier (Grenoble) **29**(3) (1979) 201–221

32. A. Unterberger, J. Unterberger, *La série discrète de* $SL(2, \mathbb{R})$ *et les opérateurs pseudo-différentiels sur une demi-droite*, Ann. Sci. Ecole Norm. Sup. **17** (1984) 83–116

33. A. Unterberger, *Symbolic calculi and the duality of homogeneous spaces*, Contemp. Math. **27** (1984) 237–252

34. A. Unterberger, J. Unterberger, *Algebras of symbols and modular forms*, J. d'Analyse Math. **68** (1996) 121–143

35. A. Unterberger, *Quantization and Non-Holomorphic Modular Forms*, Lecture Notes in Mathematics **1742**, Springer, Berlin Heidelberg New York, 2000

36. A. Unterberger, *Automorphic Pseudodifferential Analysis and Higher-Level Weyl Calculi*, Progress in Mathematics, Birkhäuser, Basel, 2002

37. A. Unterberger, *A spectral analysis of automorphic distributions and Poisson formulas*, Ann. Inst. Fourier **54**(5) (2004) 1151–1196

38. A. Unterberger, *The Fourfold Way in Real Analysis: An Alternative to the Metaplectic Representation*, Progress in Mathematics, Birkhäuser, Basel, 2006

39. N.Ya. Vilenkin, *Special Functions and Theory of Group Representations*, Translations of Mathematical Monographs, American Mathematical Society **22**, 1968

40. G. Voronoï, *Sur le développement, à l'aide des fonctions cylindriques, des sommes doubles* $\sum f(pm^2 + 2qmn + rn^2)$, Ver. Math. Kongr. Heidelberg (1904) 241–245

41. A. Weil, *Sur certains groupes d'opérateurs unitaires*, Acta Math. **111** (1964) 143–211

42. H. Weyl, *Gruppentheorie und Quantenmechanik*, Reprint of 2nd edition, Wissenschaftliche Buchgesellschaft, Darmstadt, 1977

Index

Index of Notation

A, A^*, 18
A_z, A_z^*, L_z, 20
$\mathfrak{A}$, 17
$\mathfrak{A}_v$, 75
Ana, 19
Ana$_v$, 79
$\mathfrak{B}$, 98
$C_m^{j,k}$, 38
$\mathcal{D}_{m+1}$, 15
$D_v(2\pi^{\frac{1}{2}}x)$, 77
$\mathfrak{D}$, 93
$\mathfrak{D}^\bullet$, 95
$\exp(-itL)$, 21
E, 89
E_ζ, E_ζ', 42
$\mathcal{E}$, 34
$\mathfrak{E}_{i\lambda}^\sharp$, 94
$f_{i,0}, f_{i,1}$, 17
$\mathcal{F}$, 12
$\mathcal{F}_{\mathrm{ana}}$, 19

γ_k, γ_k^*, 22
Γ_2, 95
Δ_m, 102
Θ_m, 14
$\hat{\pi}_{\rho,\varepsilon}$, 25

$\mathcal{F}_{\mathrm{ana}}^v$, 78
$F_m^{j,k}$, 39
g_z, 20, 81
$G^{(N)}$, 79
$\mathcal{G}_m(\mathbb{R}^2)$, 63
$\mathcal{H}_{m+1}$, 14
Int, 19, 78
$(\mathcal{K}u)_0, (\mathcal{K}u)_1$, 24, 76
$\mathcal{K}_{m+1}^{m_1+1,m_2+1}(\chi^1, \chi^2)$, 67, 68
$\widetilde{\mathcal{K}}_{m_1+m_2-2p}^{m_1,m_2}(\chi^1, \chi^2)$, 111
L_m, R_m, 109
$\mathcal{L}_{m+1}^{m_1+1,m_2+1}(\chi^1, \chi^2)$, 67
L, 18
$L_m^2(\mathbb{R}^2)$, 14
$\mathrm{mad}(P \wedge Q)$, 34
$\mathrm{Met}^{(n)}$, 12
$\mathrm{Met}^{(2)}$, 13
$\mathcal{N}^{j,k}(z; \mathfrak{S})$, 107

Π, 4, 20
$\tau_{y,\eta}$, 2
ϕ, 17
ϕ^j, 18
ϕ_z^j, 20

$\mathcal{N}_{\mathrm{asc}}^{j,k}(z; h)$, 103
$\mathcal{N}_m^{j,k}(z; \chi)$, 104
$\mathrm{Op}(\mathfrak{S})$, 1
$\mathrm{Op}^{\mathrm{asc}}(h), \mathrm{Op}_m^{\mathrm{asc}}(h_m)$, 30
P, Q, 17
$(Q+iP)^{-1}$, 57, 83
$(Q-\bar{z}P)^{-2}$, 59
$(\mathcal{Q}u)_0, (\mathcal{Q}u)_1$, 24, 76
$\mathcal{R}$, 34
$\mathcal{S}^A(\mathbb{R}^2)$, 28
$\mathcal{S}_m^A(\mathbb{R}^2)$, 29
$(\mathcal{S}(\mathbb{R}^2))^\dagger, (\mathcal{S}^A(\mathbb{R}^2))^\dagger$, 29
$(\mathcal{S}_{\mathrm{weak}}(\mathbb{R}^2))^\dagger$,
$\quad (\mathcal{S}_{\mathrm{weak}}'(\mathbb{R}^2))^\dagger$, 42
$\mathrm{Sp}(n,\mathbb{R}), \widetilde{\mathrm{Sp}}(n,\mathbb{R})$, 11
$\mathrm{Sq}_2(n)$, 96
$T_{\alpha,\beta}$, 28
$T_{X,s}^{j,k}, T_s^{j,k}$, 46
u_z^j, 107

$\phi_z^{v,k}$, 81
χ_{m+1}, 47
χ^v, 77
ψ^v, 77
$\#$, 27, 64

Subject Index

alternative point of view, 5
 ... pseudodifferential analysis, 27
anaplectic analysis, 16
 ... Fourier transformation, 19
 ... representation, 19
 ... rep. and Heisenberg's, 19
 ... rep. and complementary series, 25
 integral in ... , 19
 infinitesimal operators of ... rep., 20
v-anaplectic analysis, 75
 ... Fourier transformation, 78
 ... representation, 79
 integral in ... , 78
0-anaplectic and metaplectic rep., 89
ascending calculus, 30
 uniqueness of ... , 35
 covariance of ... , 33
 main theorem of ... , 46, 51
ascending–descending, 50
commuting with Q, P, 32
composition formula
 ... in anaplectic analysis, 70
 ... in Weyl calculus, 106
Dirac's comb, 93
discrete series of $SL(2, \mathbb{R})$, 15
Eisenstein distributions, 94
expansion in $\mathfrak{A}$ w.r.t. the ϕ^j's, 24
harmonic oscillator, 18, 20
Hecke's (or Bochner's) formula, 14
Heisenberg representation, 2
isotypic subspaces, 14, 28
explicit integral kernel of $\mathrm{Op}_1^{\mathrm{asc}}(h)$, 61
 ... in v-anaplectic analysis, 85
lowering operator, 18
metaplectic representation, 12
modular forms
 ... and ascending calculus, 102

nonholomorphic... and Weyl calculus,
 94, 105, 107
nonholomorphic...with weight, 104,
 110
mixed adjoint, 34
Moyal brackets, 111
nice function, 17
Poincaré's series, 99
pseudoscalar product, 18
 ... in v-anaplectic analysis, 80
quadratic transform, 12, 24, 76
quantization, 74
raising operator, 18
Rankin–Cohen brackets, 67
 ... and Moyal brackets, 114
 mock- ... , 111
 nonholomorphic ... , 106
$\mathbb{C}^4$- realization, 17
 ... in v-anaplectic analysis, 76
resolvent of lowering operator, 57
 ... in v-anaplectic analysis, 83
 components of $\mathbb{C}^4$-realization ... , 57
sharp product, 8, 27, 64, 111
spectrum of harmonic oscillator
 ... in anaplectic analysis, 18
 ... in v-anaplectic analysis, 77
star products, 74
symbol of A_z^{-m-1}, 70
symplectic group, 11
 embedding $SL(2, \mathbb{R})$ into ... , 13
$\mathcal{Q}$-transform, $\mathcal{K}$-transform, 24
 ... in v-anaplectic analysis, 76
Voronoi's identity, 98
Weyl calculus, 1
 automorphic ... , 106
 covariance of ... , 2

Lecture Notes in Mathematics

For information about earlier volumes
please contact your bookseller or Springer
LNM Online archive: springerlink.com

Vol. 1774: V. Runde, Lectures on Amenability (2002)

Vol. 1775: W. H. Meeks, A. Ros, H. Rosenberg, The Global Theory of Minimal Surfaces in Flat Spaces. Martina Franca 1999. Editor: G. P. Pirola (2002)

Vol. 1776: K. Behrend, C. Gomez, V. Tarasov, G. Tian, Quantum Comohology. Cetraro 1997. Editors: P. de Bartolomeis, B. Dubrovin, C. Reina (2002)

Vol. 1777: E. García-Río, D. N. Kupeli, R. Vázquez-Lorenzo, Osserman Manifolds in Semi-Riemannian Geometry (2002)

Vol. 1778: H. Kiechle, Theory of K-Loops (2002)

Vol. 1779: I. Chueshov, Monotone Random Systems (2002)

Vol. 1780: J. H. Bruinier, Borcherds Products on O(2,1) and Chern Classes of Heegner Divisors (2002)

Vol. 1781: E. Bolthausen, E. Perkins, A. van der Vaart, Lectures on Probability Theory and Statistics. Ecole d' Eté de Probabilités de Saint-Flour XXIX-1999. Editor: P. Bernard (2002)

Vol. 1782: C.-H. Chu, A. T.-M. Lau, Harmonic Functions on Groups and Fourier Algebras (2002)

Vol. 1783: L. Grüne, Asymptotic Behavior of Dynamical and Control Systems under Perturbation and Discretization (2002)

Vol. 1784: L. H. Eliasson, S. B. Kuksin, S. Marmi, J.-C. Yoccoz, Dynamical Systems and Small Divisors. Cetraro, Italy 1998. Editors: S. Marmi, J.-C. Yoccoz (2002)

Vol. 1785: J. Arias de Reyna, Pointwise Convergence of Fourier Series (2002)

Vol. 1786: S. D. Cutkosky, Monomialization of Morphisms from 3-Folds to Surfaces (2002)

Vol. 1787: S. Caenepeel, G. Militaru, S. Zhu, Frobenius and Separable Functors for Generalized Module Categories and Nonlinear Equations (2002)

Vol. 1788: A. Vasil'ev, Moduli of Families of Curves for Conformal and Quasiconformal Mappings (2002)

Vol. 1789: Y. Sommerhäuser, Yetter-Drinfel'd Hopf algebras over groups of prime order (2002)

Vol. 1790: X. Zhan, Matrix Inequalities (2002)

Vol. 1791: M. Knebusch, D. Zhang, Manis Valuations and Prüfer Extensions I: A new Chapter in Commutative Algebra (2002)

Vol. 1792: D. D. Ang, R. Gorenflo, V. K. Le, D. D. Trong, Moment Theory and Some Inverse Problems in Potential Theory and Heat Conduction (2002)

Vol. 1793: J. Cortés Monforte, Geometric, Control and Numerical Aspects of Nonholonomic Systems (2002)

Vol. 1794: N. Pytheas Fogg, Substitution in Dynamics, Arithmetics and Combinatorics. Editors: V. Berthé, S. Ferenczi, C. Mauduit, A. Siegel (2002)

Vol. 1795: H. Li, Filtered-Graded Transfer in Using Noncommutative Gröbner Bases (2002)

Vol. 1796: J.M. Melenk, hp-Finite Element Methods for Singular Perturbations (2002)

Vol. 1797: B. Schmidt, Characters and Cyclotomic Fields in Finite Geometry (2002)

Vol. 1798: W.M. Oliva, Geometric Mechanics (2002)

Vol. 1799: H. Pajot, Analytic Capacity, Rectifiability, Menger Curvature and the Cauchy Integral (2002)

Vol. 1800: O. Gabber, L. Ramero, Almost Ring Theory (2003)

Vol. 1801: J. Azéma, M. Émery, M. Ledoux, M. Yor (Eds.), Séminaire de Probabilités XXXVI (2003)

Vol. 1802: V. Capasso, E. Merzbach, B. G. Ivanoff, M. Dozzi, R. Dalang, T. Mountford, Topics in Spatial Stochastic Processes. Martina Franca, Italy 2001. Editor: E. Merzbach (2003)

Vol. 1803: G. Dolzmann, Variational Methods for Crystalline Microstructure – Analysis and Computation (2003)

Vol. 1804: I. Cherednik, Ya. Markov, R. Howe, G. Lusztig, Iwahori-Hecke Algebras and their Representation Theory. Martina Franca, Italy 1999. Editors: V. Baldoni, D. Barbasch (2003)

Vol. 1805: F. Cao, Geometric Curve Evolution and Image Processing (2003)

Vol. 1806: H. Broer, I. Hoveijn. G. Lunther, G. Vegter, Bifurcations in Hamiltonian Systems. Computing Singularities by Gröbner Bases (2003)

Vol. 1807: V. D. Milman, G. Schechtman (Eds.), Geometric Aspects of Functional Analysis. Israel Seminar 2000-2002 (2003)

Vol. 1808: W. Schindler, Measures with Symmetry Properties (2003)

Vol. 1809: O. Steinbach, Stability Estimates for Hybrid Coupled Domain Decomposition Methods (2003)

Vol. 1810: J. Wengenroth, Derived Functors in Functional Analysis (2003)

Vol. 1811: J. Stevens, Deformations of Singularities (2003)

Vol. 1812: L. Ambrosio, K. Deckelnick, G. Dziuk, M. Mimura, V. A. Solonnikov, H. M. Soner, Mathematical Aspects of Evolving Interfaces. Madeira, Funchal, Portugal 2000. Editors: P. Colli, J. F. Rodrigues (2003)

Vol. 1813: L. Ambrosio, L. A. Caffarelli, Y. Brenier, G. Buttazzo, C. Villani, Optimal Transportation and its Applications. Martina Franca, Italy 2001. Editors: L. A. Caffarelli, S. Salsa (2003)

Vol. 1814: P. Bank, F. Baudoin, H. Föllmer, L.C.G. Rogers, M. Soner, N. Touzi, Paris-Princeton Lectures on Mathematical Finance 2002 (2003)

Vol. 1815: A. M. Vershik (Ed.), Asymptotic Combinatorics with Applications to Mathematical Physics. St. Petersburg, Russia 2001 (2003)

Vol. 1816: S. Albeverio, W. Schachermayer, M. Talagrand, Lectures on Probability Theory and Statistics. Ecole d'Eté de Probabilités de Saint-Flour XXX-2000. Editor: P. Bernard (2003)

Vol. 1817: E. Koelink, W. Van Assche (Eds.), Orthogonal Polynomials and Special Functions. Leuven 2002 (2003)

Vol. 1818: M. Bildhauer, Convex Variational Problems with Linear, nearly Linear and/or Anisotropic Growth Conditions (2003)

Vol. 1819: D. Masser, Yu. V. Nesterenko, H. P. Schlickewei, W. M. Schmidt, M. Waldschmidt, Diophantine Approximation. Cetraro, Italy 2000. Editors: F. Amoroso, U. Zannier (2003)

Vol. 1820: F. Hiai, H. Kosaki, Means of Hilbert Space Operators (2003)

Vol. 1821: S. Teufel, Adiabatic Perturbation Theory in Quantum Dynamics (2003)

Vol. 1822: S.-N. Chow, R. Conti, R. Johnson, J. Mallet-Paret, R. Nussbaum, Dynamical Systems. Cetraro, Italy 2000. Editors: J. W. Macki, P. Zecca (2003)

Vol. 1823: A. M. Anile, W. Allegretto, C. Ringhofer, Mathematical Problems in Semiconductor Physics. Cetraro, Italy 1998. Editor: A. M. Anile (2003)

Vol. 1824: J. A. Navarro González, J. B. Sancho de Salas, $\mathscr{C}^{\infty}$ – Differentiable Spaces (2003)

Vol. 1825: J. H. Bramble, A. Cohen, W. Dahmen, Multiscale Problems and Methods in Numerical Simulations, Martina Franca, Italy 2001. Editor: C. Canuto (2003)

Vol. 1826: K. Dohmen, Improved Bonferroni Inequalities via Abstract Tubes. Inequalities and Identities of Inclusion-Exclusion Type. VIII, 113 p, 2003.

Vol. 1827: K. M. Pilgrim, Combinations of Complex Dynamical Systems. IX, 118 p, 2003.

Vol. 1828: D. J. Green, Gröbner Bases and the Computation of Group Cohomology. XII, 138 p, 2003.

Vol. 1829: E. Altman, B. Gaujal, A. Hordijk, Discrete-Event Control of Stochastic Networks: Multimodularity and Regularity. XIV, 313 p, 2003.

Vol. 1830: M. I. Gil', Operator Functions and Localization of Spectra. XIV, 256 p, 2003.

Vol. 1831: A. Connes, J. Cuntz, E. Guentner, N. Higson, J. E. Kaminker, Noncommutative Geometry, Martina Franca, Italy 2002. Editors: S. Doplicher, L. Longo (2004)

Vol. 1832: J. Azéma, M. Émery, M. Ledoux, M. Yor (Eds.), Séminaire de Probabilités XXXVII (2003)

Vol. 1833: D.-Q. Jiang, M. Qian, M.-P. Qian, Mathematical Theory of Nonequilibrium Steady States. On the Frontier of Probability and Dynamical Systems. IX, 280 p, 2004.

Vol. 1834: Yo. Yomdin, G. Comte, Tame Geometry with Application in Smooth Analysis. VIII, 186 p, 2004.

Vol. 1835: O.T. Izhboldin, B. Kahn, N.A. Karpenko, A. Vishik, Geometric Methods in the Algebraic Theory of Quadratic Forms. Summer School, Lens, 2000. Editor: J.-P. Tignol (2004)

Vol. 1836: C. Năstăsescu, F. Van Oystaeyen, Methods of Graded Rings. XIII, 304 p, 2004.

Vol. 1837: S. Tavaré, O. Zeitouni, Lectures on Probability Theory and Statistics. Ecole d'Eté de Probabilités de Saint-Flour XXXI-2001. Editor: J. Picard (2004)

Vol. 1838: A.J. Ganesh, N.W. O'Connell, D.J. Wischik, Big Queues. XII, 254 p, 2004.

Vol. 1839: R. Gohm, Noncommutative Stationary Processes. VIII, 170 p, 2004.

Vol. 1840: B. Tsirelson, W. Werner, Lectures on Probability Theory and Statistics. Ecole d'Eté de Probabilités de Saint-Flour XXXII-2002. Editor: J. Picard (2004)

Vol. 1841: W. Reichel, Uniqueness Theorems for Variational Problems by the Method of Transformation Groups (2004)

Vol. 1842: T. Johnsen, A. L. Knutsen, K_3 Projective Models in Scrolls (2004)

Vol. 1843: B. Jefferies, Spectral Properties of Noncommuting Operators (2004)

Vol. 1844: K.F. Siburg, The Principle of Least Action in Geometry and Dynamics (2004)

Vol. 1845: Min Ho Lee, Mixed Automorphic Forms, Torus Bundles, and Jacobi Forms (2004)

Vol. 1846: H. Ammari, H. Kang, Reconstruction of Small Inhomogeneities from Boundary Measurements (2004)

Vol. 1847: T.R. Bielecki, T. Björk, M. Jeanblanc, M. Rutkowski, J.A. Scheinkman, W. Xiong, Paris-Princeton Lectures on Mathematical Finance 2003 (2004)

Vol. 1848: M. Abate, J. E. Fornaess, X. Huang, J. P. Rosay, A. Tumanov, Real Methods in Complex and CR Geometry, Martina Franca, Italy 2002. Editors: D. Zaitsev, G. Zampieri (2004)

Vol. 1849: Martin L. Brown, Heegner Modules and Elliptic Curves (2004)

Vol. 1850: V. D. Milman, G. Schechtman (Eds.), Geometric Aspects of Functional Analysis. Israel Seminar 2002-2003 (2004)

Vol. 1851: O. Catoni, Statistical Learning Theory and Stochastic Optimization (2004)

Vol. 1852: A.S. Kechris, B.D. Miller, Topics in Orbit Equivalence (2004)

Vol. 1853: Ch. Favre, M. Jonsson, The Valuative Tree (2004)

Vol. 1854: O. Saeki, Topology of Singular Fibers of Differential Maps (2004)

Vol. 1855: G. Da Prato, P.C. Kunstmann, I. Lasiecka, A. Lunardi, R. Schnaubelt, L. Weis, Functional Analytic Methods for Evolution Equations. Editors: M. Iannelli, R. Nagel, S. Piazzera (2004)

Vol. 1856: K. Back, T.R. Bielecki, C. Hipp, S. Peng, W. Schachermayer, Stochastic Methods in Finance, Bressanone/Brixen, Italy, 2003. Editors: M. Fritelli, W. Runggaldier (2004)

Vol. 1857: M. Émery, M. Ledoux, M. Yor (Eds.), Séminaire de Probabilités XXXVIII (2005)

Vol. 1858: A.S. Cherny, H.-J. Engelbert, Singular Stochastic Differential Equations (2005)

Vol. 1859: E. Letellier, Fourier Transforms of Invariant Functions on Finite Reductive Lie Algebras (2005)

Vol. 1860: A. Borisyuk, G.B. Ermentrout, A. Friedman, D. Terman, Tutorials in Mathematical Biosciences I. Mathematical Neurosciences (2005)

Vol. 1861: G. Benettin, J. Henrard, S. Kuksin, Hamiltonian Dynamics – Theory and Applications, Cetraro, Italy, 1999. Editor: A. Giorgilli (2005)

Vol. 1862: B. Helffer, F. Nier, Hypoelliptic Estimates and Spectral Theory for Fokker-Planck Operators and Witten Laplacians (2005)

Vol. 1863: H. Führ, Abstract Harmonic Analysis of Continuous Wavelet Transforms (2005)

Vol. 1864: K. Efstathiou, Metamorphoses of Hamiltonian Systems with Symmetries (2005)

Vol. 1865: D. Applebaum, B.V. R. Bhat, J. Kustermans, J. M. Lindsay, Quantum Independent Increment Processes I. From Classical Probability to Quantum Stochastic Calculus. Editors: M. Schürmann, U. Franz (2005)

Vol. 1866: O.E. Barndorff-Nielsen, U. Franz, R. Gohm, B. Kümmerer, S. Thorbjønsen, Quantum Independent Increment Processes II. Structure of Quantum Lévy Processes, Classical Probability, and Physics. Editors: M. Schürmann, U. Franz, (2005)

Vol. 1867: J. Sneyd (Ed.), Tutorials in Mathematical Biosciences II. Mathematical Modeling of Calcium Dynamics and Signal Transduction. (2005)

Vol. 1868: J. Jorgenson, S. Lang, $Pos_n(R)$ and Eisenstein Series. (2005)

Vol. 1869: A. Dembo, T. Funaki, Lectures on Probability Theory and Statistics. Ecole d'Eté de Probabilités de Saint-Flour XXXIII-2003. Editor: J. Picard (2005)

Vol. 1870: V.I. Gurariy, W. Lusky, Geometry of Müntz Spaces and Related Questions. (2005)

Vol. 1871: P. Constantin, G. Gallavotti, A.V. Kazhikhov, Y. Meyer, S. Ukai, Mathematical Foundation of Turbulent Viscous Flows, Martina Franca, Italy, 2003. Editors: M. Cannone, T. Miyakawa (2006)

Vol. 1872: A. Friedman (Ed.), Tutorials in Mathematical Biosciences III. Cell Cycle, Proliferation, and Cancer (2006)

Vol. 1873: R. Mansuy, M. Yor, Random Times and Enlargements of Filtrations in a Brownian Setting (2006)

Vol. 1874: M. Yor, M. Émery (Eds.), In Memoriam Paul-André Meyer - Séminaire de Probabilités XXXIX (2006)

Vol. 1875: J. Pitman, Combinatorial Stochastic Processes. Ecole d'Eté de Probabilités de Saint-Flour XXXII-2002. Editor: J. Picard (2006)

Vol. 1876: H. Herrlich, Axiom of Choice (2006)

Vol. 1877: J. Steuding, Value Distributions of L-Functions (2007)

Vol. 1878: R. Cerf, The Wulff Crystal in Ising and Percolation Models, Ecole d'Eté de Probabilités de Saint-Flour XXXIV-2004. Editor: Jean Picard (2006)

Vol. 1879: G. Slade, The Lace Expansion and its Applications, Ecole d'Eté de Probabilités de Saint-Flour XXXIV-2004. Editor: Jean Picard (2006)

Vol. 1880: S. Attal, A. Joye, C.-A. Pillet, Open Quantum Systems I, The Hamiltonian Approach (2006)

Vol. 1881: S. Attal, A. Joye, C.-A. Pillet, Open Quantum Systems II, The Markovian Approach (2006)

Vol. 1882: S. Attal, A. Joye, C.-A. Pillet, Open Quantum Systems III, Recent Developments (2006)

Vol. 1883: W. Van Assche, F. Marcellàn (Eds.), Orthogonal Polynomials and Special Functions, Computation and Application (2006)

Vol. 1884: N. Hayashi, E.I. Kaikina, P.I. Naumkin, I.A. Shishmarev, Asymptotics for Dissipative Nonlinear Equations (2006)

Vol. 1885: A. Telcs, The Art of Random Walks (2006)

Vol. 1886: S. Takamura, Splitting Deformations of Degenerations of Complex Curves (2006)

Vol. 1887: K. Habermann, L. Habermann, Introduction to Symplectic Dirac Operators (2006)

Vol. 1888: J. van der Hoeven, Transseries and Real Differential Algebra (2006)

Vol. 1889: G. Osipenko, Dynamical Systems, Graphs, and Algorithms (2006)

Vol. 1890: M. Bunge, J. Funk, Singular Coverings of Toposes (2006)

Vol. 1891: J.B. Friedlander, D.R. Heath-Brown, H. Iwaniec, J. Kaczorowski, Analytic Number Theory, Cetraro, Italy, 2002. Editors: A. Perelli, C. Viola (2006)

Vol. 1892: A. Baddeley, I. Bárány, R. Schneider, W. Weil, Stochastic Geometry, Martina Franca, Italy, 2004. Editor: W. Weil (2007)

Vol. 1893: H. Hanßmann, Local and Semi-Local Bifurcations in Hamiltonian Dynamical Systems, Results and Examples (2007)

Vol. 1894: C.W. Groetsch, Stable Approximate Evaluation of Unbounded Operators (2007)

Vol. 1895: L. Molnár, Selected Preserver Problems on Algebraic Structures of Linear Operators and on Function Spaces (2007)

Vol. 1896: P. Massart, Concentration Inequalities and Model Selection, Ecole d'Été de Probabilités de Saint-Flour XXXIII-2003. Editor: J. Picard (2007)

Vol. 1897: R. Doney, Fluctuation Theory for Lévy Processes, Ecole d'Été de Probabilités de Saint-Flour XXXV-2005. Editor: J. Picard (2007)

Vol. 1898: H.R. Beyer, Beyond Partial Differential Equations, On linear and Quasi-Linear Abstract Hyperbolic Evolution Equations (2007)

Vol. 1899: Séminaire de Probabilités XL. Editors: C. Donati-Martin, M. Émery, A. Rouault, C. Stricker (2007)

Vol. 1900: E. Bolthausen, A. Bovier (Eds.), Spin Glasses (2007)

Vol. 1901: O. Wittenberg, Intersections de deux quadriques et pinceaux de courbes de genre 1, Intersections of Two Quadrics and Pencils of Curves of Genus 1 (2007)

Vol. 1902: A. Isaev, Lectures on the Automorphism Groups of Kobayashi-Hyperbolic Manifolds (2007)

Vol. 1903: G. Kresin, V. Maz'ya, Sharp Real-Part Theorems (2007)

Vol. 1904: P. Giesl, Construction of Global Lyapunov Functions Using Radial Basis Functions (2007)

Vol. 1905: C. Prévôt, M. Röckner, A Concise Course on Stochastic Partial Differential Equations (2007)

Vol. 1906: T. Schuster, The Method of Approximate Inverse: Theory and Applications (2007)

Vol. 1907: M. Rasmussen, Attractivity and Bifurcation for Nonautonomous Dynamical Systems (2007)

Vol. 1908: T.J. Lyons, M. Caruana, T. Lévy, Differential Equations Driven by Rough Paths, Ecole d'Été de Probabilités de Saint-Flour XXXIV-2004 (2007)

Vol. 1909: H. Akiyoshi, M. Sakuma, M. Wada, Y. Yamashita, Punctured Torus Groups and 2-Bridge Knot Groups (I) (2007)

Vol. 1910: V.D. Milman, G. Schechtman (Eds.), Geometric Aspects of Functional Analysis. Israel Seminar 2004-2005 (2007)

Vol. 1911: A. Bressan, D. Serre, M. Williams, K. Zumbrun, Hyperbolic Systems of Balance Laws. Cetraro, Italy 2003. Editor: P. Marcati (2007)

Vol. 1912: V. Berinde, Iterative Approximation of Fixed Points (2007)

Vol. 1913: J.E. Marsden, G. Misiołek, J.-P. Ortega, M. Perlmutter, T.S. Ratiu, Hamiltonian Reduction by Stages (2007)

Vol. 1914: G. Kutyniok, Affine Density in Wavelet Analysis (2007)

Vol. 1915: T. Bıyıkoğlu, J. Leydold, P.F. Stadler, Laplacian Eigenvectors of Graphs. Perron-Frobenius and Faber-Krahn Type Theorems (2007)

Vol. 1916: C. Villani, F. Rezakhanlou, Entropy Methods for the Boltzmann Equation. Editors: F. Golse, S. Olla (2008)

Vol. 1917: I. Veselić, Existence and Regularity Properties of the Integrated Density of States of Random Schrödinger (2008)

Vol. 1918: B. Roberts, R. Schmidt, Local Newforms for GSp(4) (2007)

Vol. 1919: R.A. Carmona, I. Ekeland, A. Kohatsu-Higa, J.-M. Lasry, P.-L. Lions, H. Pham, E. Taflin, Paris-Princeton Lectures on Mathematical Finance 2004.

Editors: R.A. Carmona, E. Çinlar, I. Ekeland, E. Jouini, J.A. Scheinkman, N. Touzi (2007)

Vol. 1920: S.N. Evans, Probability and Real Trees. Ecole d'Été de Probabilités de Saint-Flour XXXV-2005 (2008)

Vol. 1921: J.P. Tian, Evolution Algebras and their Applications (2008)

Vol. 1922: A. Friedman (Ed.), Tutorials in Mathematical BioSciences IV. Evolution and Ecology (2008)

Vol. 1923: J.P.N. Bishwal, Parameter Estimation in Stochastic Differential Equations (2008)

Vol. 1924: M. Wilson, Littlewood-Paley Theory and Exponential-Square Integrability (2008)

Vol. 1925: M. du Sautoy, L. Woodward, Zeta Functions of Groups and Rings (2008)

Vol. 1926: L. Barreira, V. Claudia, Stability of Nonautonomous Differential Equations (2008)

Vol. 1927: L. Ambrosio, L. Caffarelli, M.G. Crandall, L.C. Evans, N. Fusco, Calculus of Variations and Non-Linear Partial Differential Equations. Cetraro, Italy 2005. Editors: B. Dacorogna, P. Marcellini (2008)

Vol. 1928: J. Jonsson, Simplicial Complexes of Graphs (2008)

Vol. 1929: Y. Mishura, Stochastic Calculus for Fractional Brownian Motion and Related Processes (2008)

Vol. 1930: J.M. Urbano, The Method of Intrinsic Scaling. A Systematic Approach to Regularity for Degenerate and Singular PDEs (2008)

Vol. 1931: M. Cowling, E. Frenkel, M. Kashiwara, A. Valette, D.A. Vogan, Jr., N.R. Wallach, Representation Theory and Complex Analysis. Venice, Italy 2004. Editors; E.C. Tarabusi, A. D'Agnolo, M. Picardello (2008)

Vol. 1932: A.A. Agrachev, A.S. Morse, E.D. Sontag, H.J. Sussmann, V.I. Utkin, Nonlinear and Optimal Control Theory. Cetraro, Italy 2004. Editors: P. Nistri, G. Stefani (2008)

Vol. 1933: M. Petkovic, Point Estimation of Root Finding Methods (2008)

Vol. 1934: C. Donati-Martin, M. Émery, A. Rouault, C. Stricker (Eds.), Séminaire de Probabilités XLI (2008)

Vol. 1935: A. Unterberger, Alternative Pseudodifferential Analysis (2008)

Vol. 1936: P. Magal, S. Ruan (Eds.), Structured Population Models in Biology and Epidemiology (2008)

Vol. 1937: G. Capriz, P. Giovine, P.M. Mariano (Eds.), Mathematical Models of Granular Matter (2008)

Vol. 1938: D. Auroux, F. Catanese, M. Manetti, P. Seidel, B. Siebert, I. Smith, G. Tian, Symplectic 4-Manifolds and Algebraic Surfaces. Cetraro, Italy 2003. Editors: F. Catanese, G. Tian (2008)

Vol. 1939: D. Boffi, F. Brezzi, L. Demkowicz, R.G. Durán, R.S. Falk, M. Fortin, Mixed Finite Elements, Compatibility Conditions, and Applications. Cetraro, Italy 2006. Editors: D. Boffi, L. Gastaldi (2008)

Vol. 1940: J. Banasiak, V. Capasso, M.A.J. Chaplain, M. Lachowicz, J. Miękisz, Multiscale Problems in the Life Sciences. From Microscopic to Macroscopic. Będlewo, Poland 2006. Editors: V. Capasso, M. Lachowicz (2008)

Vol. 1941: S.M.J. Haran, Arithmetical Investigations. Representation Theory, Orthogonal Polynomials, and Quantum Interpolations (2008)

Vol. 1942: S. Albeverio, F. Flandoli, Y.G. Sinai, SPDE in Hydrodynamic. Recent Progress and Prospects. Cetraro, Italy 2005. Editors: G. Da Prato, M. Röckner (2008)

Vol. 1943: L.L. Bonilla (Ed.), Inverse Problems and Imaging. Martina Franca, Italy 2002 (2008)

Vol. 1944: A. Di Bartolo, G. Falcone, P. Plaumann, K. Strambach, Algebraic Groups and Lie Groups with Few Factors (2008)

Vol. 1945: F. Brauer, P. van den Driessche, J. Wu (Eds.), Mathematical Epidemiology (2008)

Vol. 1946: G. Allaire, A. Arnold, P. Degond, T.Y. Hou, Quantum Transport. Modelling, Analysis and Asymptotics. Cetraro, Italy 2006. Editors: N.B. Abdallah, G. Frosali (2008)

Vol. 1947: D. Abramovich, M. Mariño, M. Thaddeus, R. Vakil, Enumerative Invariants in Algebraic Geometry and String Theory. Cetraro, Italy 2005. Editors: K. Behrend, M. Manetti (2008)

Vol. 1948: F. Cao, J-L. Lisani, J-M. Morel, P. Musé, F. Sur, A Theory of Shape Identification (2008)

Vol. 1949: H.G. Feichtinger, B. Helffer, M.P. Lamoureux, N. Lerner, J. Toft, Pseudo-Differential Operators. Quantization and Signals. Cetraro, Italy 2006. Editors: L. Rodino, M.W. Wong (2008)

Vol. 1950: M. Bramson, Stability of Queueing Networks, Ecole d'Eté de Probabilités de Saint-Flour XXXVI-2006 (2008)

Vol. 1951: A. Moltó, J. Orihuela, S. Troyanski, M. Valdivia, A Non Linear Transfer Technique for Renorming (2008)

Vol. 1952: R. Mikhailov, I.B.S. Passi, Lower Central and Dimension Series of Groups (2008)

Vol. 1953: K. Arwini, C.T.J. Dodson, Information Geometry (2008)

Vol. 1954: P. Biane, L. Bouten, F. Cipriani, N. Konno, N. Privault, Q. Xu, Quantum Potential Theory. Editors: U. Franz, M. Schuermann (2008)

Vol. 1955: M. Bernot, V. Caselles, J.-M. Morel, Optimal transportation networks (2008)

Vol. 1956: C.H. Chu, Matrix Convolution Operators on Groups (2008)

Vol. 1957: A. Guionnet, On Random Matrices: Macroscopic Asymptotics, Ecole d'Eté de Probabilités de Saint-Flour XXXVI-2006 (2008)

Vol. 1958: M.C. Olsson, Compactifying Moduli Spaces for Abelian Varieties (2008)

Recent Reprints and New Editions

Vol. 1702: J. Ma, J. Yong, Forward-Backward Stochastic Differential Equations and their Applications. 1999 – Corr. 3rd printing (2007)

Vol. 830: J.A. Green, Polynomial Representations of GL_n, with an Appendix on Schensted Correspondence and Littelmann Paths by K. Erdmann, J.A. Green and M. Schoker 1980 – 2nd corr. and augmented edition (2007)

Vol. 1693: S. Simons, From Hahn-Banach to Monotonicity (Minimax and Monotonicity 1998) – 2nd exp. edition (2008)

Vol. 470: R.E. Bowen, Equilibrium States and the Ergodic Theory of Anosov Diffeomorphisms. With a preface by D. Ruelle. Edited by J.-R. Chazottes. 1975 – 2nd rev. edition (2008)

Vol. 523: S.A. Albeverio, R.J. Høegh-Krohn, S. Mazzucchi, Mathematical Theory of Feynman Path Integral. 1976 – 2nd corr. and enlarged edition (2008)

Vol. 1764: A. Cannas da Silva, Lectures on Symplectic Geometry 2001 – Corr. 2nd printing (2008)